FREE Study Skills Videos/DVD Offer

Dear Customer,

Thank you for your purchase from Mometrix! We consider it an honor and a privilege that you have purchased our product and we want to ensure your satisfaction.

As part of our ongoing effort to meet the needs of test takers, we have developed a set of Study Skills Videos that we would like to give you for <u>FREE</u>. These videos cover our *best practices* for getting ready for your exam, from how to use our study materials to how to best prepare for the day of the test.

All that we ask is that you email us with feedback that would describe your experience so far with our product. Good, bad, or indifferent, we want to know what you think!

To get your FREE Study Skills Videos, you can use the **QR code** below, or send us an **email** at studyvideos@mometrix.com with *FREE VIDEOS* in the subject line and the following information in the body of the email:

- The name of the product you purchased.
- Your product rating on a scale of 1-5, with 5 being the highest rating.
- Your feedback. It can be long, short, or anything in between. We just want to know your impressions and experience so far with our product. (Good feedback might include how our study material met your needs and ways we might be able to make it even better. You could highlight features that you found helpful or features that you think we should add.)

If you have any questions or concerns, please don't hesitate to contact me directly.

Thanks again!

Sincerely,

Jay Willis
Vice President
jay.willis@mometrix.com
1-800-673-8175

Journeyman Electrician

Exam Prep Practice Questions

Full-Length Tests Based on the NEC 2023 National Electrical Code Book

2nd Edition

Written and edited by Mometrix Test Prep

Printed in the United States of America

This paper meets the requirements of ANSI/NISO Z39.48-1992 (Permanence of Paper).

Mometrix offers volume discount pricing to institutions. For more information or a price quote, please contact our sales department at sales@mometrix.com or 888-248-1219.

Mometrix Media LLC is not affiliated with or endorsed by any official testing organization. All organizational and test names are trademarks of their respective owners.

Paperback
ISBN 13: 978-1-5167-2314-0
ISBN 10: 1-5167-2314-7

DEAR FUTURE EXAM SUCCESS STORY

First of all, **THANK YOU** for purchasing Mometrix study materials!

Second, congratulations! You are one of the few determined test-takers who are committed to doing whatever it takes to excel on your exam. **You have come to the right place.** We developed these study materials with one goal in mind: to deliver you the information you need in a format that's concise and easy to use.

In addition to optimizing your guide for the content of the test, we've outlined our recommended steps for breaking down the preparation process into small, attainable goals so you can make sure you stay on track.

We've also analyzed the entire test-taking process, identifying the most common pitfalls and showing how you can overcome them and be ready for any curveball the test throws you.

Standardized testing is one of the biggest obstacles on your road to success, which only increases the importance of doing well in the high-pressure, high-stakes environment of test day. Your results on this test could have a significant impact on your future, and this guide provides the information and practical advice to help you achieve your full potential on test day.

Your success is our success

We would love to hear from you! If you would like to share the story of your exam success or if you have any questions or comments in regard to our products, please contact us at **800-673-8175** or **support@mometrix.com**.

Thanks again for your business and we wish you continued success!

Sincerely,
The Mometrix Test Preparation Team

TABLE OF CONTENTS

Introduction

Thank you for purchasing this resource! You have made the choice to prepare yourself for a test that could have a huge impact on your future, and this guide is designed to help you be fully ready for test day. Obviously, it's important to have a solid understanding of the test material, but you also need to be prepared for the unique environment and stressors of the test, so that you can perform to the best of your abilities.

For this purpose, the first section that appears in this guide is the **Secret Keys**. We've devoted countless hours to meticulously researching what works and what doesn't, and we've boiled down our findings to the five most impactful steps you can take to improve your performance on the test. We start at the beginning with study planning and move through the preparation process, all the way to the testing strategies that will help you get the most out of what you know when you're finally sitting in front of the test.

We recommend that you start preparing for your test as far in advance as possible. However, if you've bought this guide as a last-minute study resource and only have a few days before your test, we recommend that you skip over the first two Secret Keys since they address a long-term study plan.

If you struggle with **test anxiety**, we strongly encourage you to check out our recommendations for how you can overcome it. Test anxiety is a formidable foe, but it can be beaten, and we want to make sure you have the tools you need to defeat it.

1

Secret Key 1: Plan Big, Study Small

There's a lot riding on your performance. If you want to ace this test, you're going to need to keep your skills sharp and the material fresh in your mind. You need a plan that lets you review everything you need to know while still fitting in your schedule. We'll break this strategy down into three categories.

Information Organization

Start with the information you already have: the official test outline. From this, you can make a complete list of all the concepts you need to cover before the test. Organize these concepts into groups that can be studied together, and create a list of any related vocabulary you need to learn so you can brush up on any difficult terms. You'll want to keep this vocabulary list handy once you actually start studying since you may need to add to it along the way.

Time Management

Once you have your set of study concepts, decide how to spread them out over the time you have left before the test. Break your study plan into small, clear goals so you have a manageable task for each day and know exactly what you're doing. Then just focus on one small step at a time. When you manage your time this way, you don't need to spend hours at a time studying. Studying a small block of content for a short period each day helps you retain information better and avoid stressing over how much you have left to do. You can relax knowing that you have a plan to cover everything in time. In order for this strategy to be effective though, you have to start studying early and stick to your schedule. Avoid the exhaustion and futility that comes from last-minute cramming!

Study Environment

The environment you study in has a big impact on your learning. Studying in a coffee shop, while probably more enjoyable, is not likely to be as fruitful as studying in a quiet room. It's important to keep distractions to a minimum. You're only planning to study for a short block of time, so make the most of it. Don't pause to check your phone or get up to find a snack. It's also important to **avoid multitasking**. Research has consistently shown that multitasking will make your studying dramatically less effective. Your study area should also be comfortable and well-lit so you don't have the distraction of straining your eyes or sitting on an uncomfortable chair.

 The time of day you study is also important. You want to be rested and alert. Don't wait until just before bedtime. Study when you'll be most likely to comprehend and remember. Even better, if you know what time of day your test will be, set that time aside for study. That way your brain will be used to working on that subject at that specific time and you'll have a better chance of recalling information.

2

Finally, it can be helpful to team up with others who are studying for the same test. Your actual studying should be done in as isolated an environment as possible, but the work of organizing the information and setting up the study plan can be divided up. In between study sessions, you can discuss with your teammates the concepts that you're all studying and quiz each other on the details. Just be sure that your teammates are as serious about the test as you are. If you find that your study time is being replaced with social time, you might need to find a new team.

Secret Key 2: Make Your Studying Count

You're devoting a lot of time and effort to preparing for this test, so you want to be absolutely certain it will pay off. This means doing more than just reading the content and hoping you can remember it on test day. It's important to make every minute of study count. There are two main areas you can focus on to make your studying count.

Retention

It doesn't matter how much time you study if you can't remember the material. You need to make sure you are retaining the concepts. To check your retention of the information you're learning, try recalling it at later times with minimal prompting. Try carrying around flashcards and glance at one or two from time to time or ask a friend who's also studying for the test to quiz you.

To enhance your retention, look for ways to put the information into practice so that you can apply it rather than simply recalling it. If you're using the information in practical ways, it will be much easier to remember. Similarly, it helps to solidify a concept in your mind if you're not only reading it to yourself but also explaining it to someone else. Ask a friend to let you teach them about a concept you're a little shaky on (or speak aloud to an imaginary audience if necessary). As you try to summarize, define, give examples, and answer your friend's questions, you'll understand the concepts better and they will stay with you longer. Finally, step back for a big picture view and ask yourself how each piece of information fits with the whole subject. When you link the different concepts together and see them working together as a whole, it's easier to remember the individual components.

Finally, practice showing your work on any multi-step problems, even if you're just studying. Writing out each step you take to solve a problem will help solidify the process in your mind, and you'll be more likely to remember it during the test.

Modality

Modality simply refers to the means or method by which you study. Choosing a study modality that fits your own individual learning style is crucial. No two people learn best in exactly the same way, so it's important to know your strengths and use them to your advantage.

For example, if you learn best by visualization, focus on visualizing a concept in your mind and draw an image or a diagram. Try color-coding your notes, illustrating them, or creating symbols that will trigger your mind to recall a learned concept. If you learn best by hearing or discussing information, find a study partner who learns the same way or read aloud to yourself. Think about how to put the information in your own words. Imagine that you are giving a lecture on the topic and record yourself so you can listen to it later.

For any learning style, flashcards can be helpful. Organize the information so you can take advantage of spare moments to review. Underline key words or phrases. Use different colors for different categories. Mnemonic devices (such as creating a short list in which every item starts with the same letter) can also help with retention. Find what works best for you and use it to store the information in your mind most effectively and easily.

Secret Key 3: Practice the Right Way

Your success on test day depends not only on how many hours you put into preparing, but also on whether you prepared the right way. It's good to check along the way to see if your studying is paying off. One of the most effective ways to do this is by taking practice tests to evaluate your progress. Practice tests are useful because they show exactly where you need to improve. Every time you take a practice test, pay special attention to these three groups of questions:

- The questions you got wrong
- The questions you had to guess on, even if you guessed right
- The questions you found difficult or slow to work through

This will show you exactly what your weak areas are, and where you need to devote more study time. Ask yourself why each of these questions gave you trouble. Was it because you didn't understand the material? Was it because you didn't remember the vocabulary? Do you need more repetitions on this type of question to build speed and confidence? Dig into those questions and figure out how you can strengthen your weak areas as you go back to review the material.

 Additionally, many practice tests have a section explaining the answer choices. It can be tempting to read the explanation and think that you now have a good understanding of the concept. However, an explanation likely only covers part of the question's broader context. Even if the explanation makes perfect sense, **go back and investigate** every concept related to the question until you're positive you have a thorough understanding.

As you go along, keep in mind that the practice test is just that: practice. Memorizing these questions and answers will not be very helpful on the actual test because it is unlikely to have any of the same exact questions. If you only know the right answers to the sample questions, you won't be prepared for the real thing. **Study the concepts** until you understand them fully, and then you'll be able to answer any question that shows up on the test.

It's important to wait on the practice tests until you're ready. If you take a test on your first day of study, you may be overwhelmed by the amount of material covered and how much you need to learn. Work up to it gradually.

On test day, you'll need to be prepared for answering questions, managing your time, and using the test-taking strategies you've learned. It's a lot to balance, like a mental marathon that will have a big impact on your future. Like training for a marathon, you'll need to start slowly and work your way up. When test day arrives, you'll be ready.

Start with the strategies you've read in the first two Secret Keys—plan your course and study in the way that works best for you. If you have time, consider using multiple study

6

resources to get different approaches to the same concepts. It can be helpful to see difficult concepts from more than one angle. Then find a good source for practice tests. Many times, the test website will suggest potential study resources or provide sample tests.

Practice Test Strategy

If you're able to find at least three practice tests, we recommend this strategy:

UNTIMED AND OPEN-BOOK PRACTICE

Take the first test with no time constraints and with your notes and study guide handy. Take your time and focus on applying the strategies you've learned.

TIMED AND OPEN-BOOK PRACTICE

Take the second practice test open-book as well, but set a timer and practice pacing yourself to finish in time.

TIMED AND CLOSED-BOOK PRACTICE

Take any other practice tests as if it were test day. Set a timer and put away your study materials. Sit at a table or desk in a quiet room, imagine yourself at the testing center, and answer questions as quickly and accurately as possible.

Keep repeating timed and closed-book tests on a regular basis until you run out of practice tests or it's time for the actual test. Your mind will be ready for the schedule and stress of test day, and you'll be able to focus on recalling the material you've learned.

Secret Key 4: Pace Yourself

Once you're fully prepared for the material on the test, your biggest challenge on test day will be managing your time. Just knowing that the clock is ticking can make you panic even if you have plenty of time left. Work on pacing yourself so you can build confidence against the time constraints of the exam. Pacing is a difficult skill to master, especially in a high-pressure environment, so **practice is vital**.

Set time expectations for your pace based on how much time is available. For example, if a section has 60 questions and the time limit is 30 minutes, you know you have to average 30 seconds or less per question in order to answer them all. Although 30 seconds is the hard limit, set 25 seconds per question as your goal, so you reserve extra time to spend on harder questions. When you budget extra time for the harder questions, you no longer have any reason to stress when those questions take longer to answer.

Don't let this time expectation distract you from working through the test at a calm, steady pace, but keep it in mind so you don't spend too much time on any one question. Recognize that taking extra time on one question you don't understand may keep you from answering two that you do understand later in the test. If your time limit for a question is up and you're still not sure of the answer, mark it and move on, and come back to it later if the time and the test format allow. If the testing format doesn't allow you to return to earlier questions, just make an educated guess; then put it out of your mind and move on.

On the easier questions, be careful not to rush. It may seem wise to hurry through them so you have more time for the challenging ones, but it's not worth missing one if you know the concept and just didn't take the time to read the question fully. Work efficiently but make sure you understand the question and have looked at all of the answer choices, since more than one may seem right at first.

Even if you're paying attention to the time, you may find yourself a little behind at some point. You should speed up to get back on track, but do so wisely. Don't panic; just take a few seconds less on each question until you're caught up. Don't guess without thinking, but do look through the answer choices and eliminate any you know are wrong. If you can get down to two choices, it is often worthwhile to guess from those. Once you've chosen an answer, move on and don't dwell on any that you skipped or had to hurry through. If a question was taking too long, chances are it was one of the harder ones, so you weren't as likely to get it right anyway.

On the other hand, if you find yourself getting ahead of schedule, it may be beneficial to slow down a little. The more quickly you work, the more likely you are to make a careless mistake that will affect your score. You've budgeted time for each question, so don't be afraid to spend that time. Practice an efficient but careful pace to get the most out of the time you have.

Secret Key 5: Have a Plan for Guessing

When you're taking the test, you may find yourself stuck on a question. Some of the answer choices seem better than others, but you don't see the one answer choice that is obviously correct. What do you do?

The scenario described above is very common, yet most test takers have not effectively prepared for it. Developing and practicing a plan for guessing may be one of the single most effective uses of your time as you get ready for the exam.

In developing your plan for guessing, there are three questions to address:

- When should you start the guessing process?
- How should you narrow down the choices?
- Which answer should you choose?

When to Start the Guessing Process

Unless your plan for guessing is to select C every time (which, despite its merits, is not what we recommend), you need to leave yourself enough time to apply your answer elimination strategies. Since you have a limited amount of time for each question, that means that if you're going to give yourself the best shot at guessing correctly, you have to decide quickly whether or not you will guess.

Of course, the best-case scenario is that you don't have to guess at all, so first, see if you can answer the question based on your knowledge of the subject and basic reasoning skills. Focus on the key words in the question and try to jog your memory of related topics. Give yourself a chance to bring the knowledge to mind, but once you realize that you don't have (or you can't access) the knowledge you need to answer the question, it's time to start the guessing process.

It's almost always better to start the guessing process too early than too late. It only takes a few seconds to remember something and answer the question from knowledge. Carefully eliminating wrong answer choices takes longer. Plus, going through the process of eliminating answer choices can actually help jog your memory.

Summary: Start the guessing process as soon as you decide that you can't answer the question based on your knowledge.

9

How to Narrow Down the Choices

The next chapter in this book (**Test-Taking Strategies**) includes a wide range of strategies for how to approach questions and how to look for answer choices to eliminate. You will definitely want to read those carefully, practice them, and figure out which ones work best for you. Here though, we're going to address a mindset rather than a particular strategy.

Your odds of guessing an answer correctly depend on how many options you are choosing from.

Number of options left	5	4	3	2	1
Odds of guessing correctly	20%	25%	33%	50%	100%

You can see from this chart just how valuable it is to be able to eliminate incorrect answers and make an educated guess, but there are two things that many test takers do that cause them to miss out on the benefits of guessing:

- Accidentally eliminating the correct answer
- Selecting an answer based on an impression

We'll look at the first one here, and the second one in the next section.

To avoid accidentally eliminating the correct answer, we recommend a thought exercise called **the $5 challenge**. In this challenge, you only eliminate an answer choice from contention if you are willing to bet $5 on it being wrong. Why $5? Five dollars is a small but not insignificant amount of money. It's an amount you could afford to lose but wouldn't

want to throw away. And while losing $5 once might not hurt too much, doing it twenty times will set you back $100. In the same way, each small decision you make—eliminating a choice here, guessing on a question there—won't by itself impact your score very much, but when you put them all together, they can make a big difference. By holding each answer choice elimination decision to a higher standard, you can reduce the risk of accidentally eliminating the correct answer.

The $5 challenge can also be applied in a positive sense: If you are willing to bet $5 that an answer choice *is* correct, go ahead and mark it as correct.

Summary: Only eliminate an answer choice if you are willing to bet $5 that it is wrong.

Which Answer to Choose

You're taking the test. You've run into a hard question and decided you'll have to guess. You've eliminated all the answer choices you're willing to bet $5 on. Now you have to pick an answer. Why do we even need to talk about this? Why can't you just pick whichever one you feel like when the time comes?

The answer to these questions is that if you don't come into the test with a plan, you'll rely on your impression to select an answer choice, and if you do that, you risk falling into a trap. The test writers know that everyone who takes their test will be guessing on some of the questions, so they intentionally write wrong answer choices to seem plausible. You still have to pick an answer though, and if the wrong answer choices are designed to look right, how can you ever be sure that you're not falling for their trap? The best solution we've found to this dilemma is to take the decision out of your hands entirely. Here is the process we recommend:

Once you've eliminated any choices that you are confident (willing to bet $5) are wrong, select the first remaining choice as your answer.

Whether you choose to select the first remaining choice, the second, or the last, the important thing is that you use some preselected standard. Using this approach guarantees that you will not be enticed into selecting an answer choice that looks right, because you are not basing your decision on how the answer choices look.

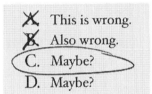

This is not meant to make you question your knowledge. Instead, it is to help you recognize the difference between your knowledge and your impressions. There's a huge difference between thinking an answer is right because of what you know, and thinking an answer is right because it looks or sounds like it should be right.

Summary: To ensure that your selection is appropriately random, make a predetermined selection from among all answer choices you have not eliminated.

Test-Taking Strategies

This section contains a list of test-taking strategies that you may find helpful as you work through the test. By taking what you know and applying logical thought, you can maximize your chances of answering any question correctly!

It is very important to realize that every question is different and every person is different: no single strategy will work on every question, and no single strategy will work for every person. That's why we've included all of them here, so you can try them out and determine which ones work best for different types of questions and which ones work best for you.

Question Strategies

⌀ READ CAREFULLY

Read the question and the answer choices carefully. Don't miss the question because you misread the terms. You have plenty of time to read each question thoroughly and make sure you understand what is being asked. Yet a happy medium must be attained, so don't waste too much time. You must read carefully and efficiently.

⌀ CONTEXTUAL CLUES

Look for contextual clues. If the question includes a word you are not familiar with, look at the immediate context for some indication of what the word might mean. Contextual clues can often give you all the information you need to decipher the meaning of an unfamiliar word. Even if you can't determine the meaning, you may be able to narrow down the possibilities enough to make a solid guess at the answer to the question.

⌀ PREFIXES

If you're having trouble with a word in the question or answer choices, try dissecting it. Take advantage of every clue that the word might include. Prefixes can be a huge help. Usually, they allow you to determine a basic meaning. *Pre-* means before, *post-* means after, *pro-* is positive, *de-* is negative. From prefixes, you can get an idea of the general meaning of the word and try to put it into context.

⌀ HEDGE WORDS

Watch out for critical hedge words, such as *likely, may, can, sometimes, often, almost, mostly, usually, generally, rarely,* and *sometimes.* Question writers insert these hedge phrases to cover every possibility. Often an answer choice will be wrong simply because it leaves no room for exception. Be on guard for answer choices that have definitive words such as *exactly* and *always.*

⊘ SWITCHBACK WORDS

Stay alert for *switchbacks*. These are the words and phrases frequently used to alert you to shifts in thought. The most common switchback words are *but, although,* and *however.* Others include *nevertheless, on the other hand, even though, while, in spite of, despite,* and *regardless of.* Switchback words are important to catch because they can change the direction of the question or an answer choice.

⊘ FACE VALUE

When in doubt, use common sense. Accept the situation in the problem at face value. Don't read too much into it. These problems will not require you to make wild assumptions. If you have to go beyond creativity and warp time or space in order to have an answer choice fit the question, then you should move on and consider the other answer choices. These are normal problems rooted in reality. The applicable relationship or explanation may not be readily apparent, but it is there for you to figure out. Use your common sense to interpret anything that isn't clear.

Answer Choice Strategies

⊘ ANSWER SELECTION

The most thorough way to pick an answer choice is to identify and eliminate wrong answers until only one is left, then confirm it is the correct answer. Sometimes an answer choice may immediately seem right, but be careful. The test writers will usually put more than one reasonable answer choice on each question, so take a second to read all of them and make sure that the other choices are not equally obvious. As long as you have time left, it is better to read every answer choice than to pick the first one that looks right without checking the others.

⊘ ANSWER CHOICE FAMILIES

An answer choice family consists of two (in rare cases, three) answer choices that are very similar in construction and cannot all be true at the same time. If you see two answer choices that are direct opposites or parallels, one of them is usually the correct answer. For instance, if one answer choice says that quantity x increases and another either says that quantity x decreases (opposite) or says that quantity y increases (parallel), then those answer choices would fall into the same family. An answer choice that doesn't match the construction of the answer choice family is more likely to be incorrect. Most questions will not have answer choice families, but when they do appear, you should be prepared to recognize them.

⊘ ELIMINATE ANSWERS

Eliminate answer choices as soon as you realize they are wrong, but make sure you consider all possibilities. If you are eliminating answer choices and realize that the last one you are left with is also wrong, don't panic. Start over and consider each choice again. There may be something you missed the first time that you will realize on the second pass.

⊘ Avoid Fact Traps

Don't be distracted by an answer choice that is factually true but doesn't answer the question. You are looking for the choice that answers the question. Stay focused on what the question is asking for so you don't accidentally pick an answer that is true but incorrect. Always go back to the question and make sure the answer choice you've selected actually answers the question and is not merely a true statement.

⊘ Extreme Statements

In general, you should avoid answers that put forth extreme actions as standard practice or proclaim controversial ideas as established fact. An answer choice that states the "process should be used in certain situations, if…" is much more likely to be correct than one that states the "process should be discontinued completely." The first is a calm rational statement and doesn't even make a definitive, uncompromising stance, using a hedge word *if* to provide wiggle room, whereas the second choice is far more extreme.

⊘ Benchmark

As you read through the answer choices and you come across one that seems to answer the question well, mentally select that answer choice. This is not your final answer, but it's the one that will help you evaluate the other answer choices. The one that you selected is your benchmark or standard for judging each of the other answer choices. Every other answer choice must be compared to your benchmark. That choice is correct until proven otherwise by another answer choice beating it. If you find a better answer, then that one becomes your new benchmark. Once you've decided that no other choice answers the question as well as your benchmark, you have your final answer.

⊘ Predict the Answer

Before you even start looking at the answer choices, it is often best to try to predict the answer. When you come up with the answer on your own, it is easier to avoid distractions and traps because you will know exactly what to look for. The right answer choice is unlikely to be word-for-word what you came up with, but it should be a close match. Even if you are confident that you have the right answer, you should still take the time to read each option before moving on.

General Strategies

⊘ Tough Questions

If you are stumped on a problem or it appears too hard or too difficult, don't waste time. Move on! Remember though, if you can quickly check for obviously incorrect answer choices, your chances of guessing correctly are greatly improved. Before you completely give up, at least try to knock out a couple of possible answers. Eliminate what you can and then guess at the remaining answer choices before moving on.

⊘ Check Your Work

Since you will probably not know every term listed and the answer to every question, it is important that you get credit for the ones that you do know. Don't miss any questions through careless mistakes. If at all possible, try to take a second to look back over your

14

answer selection and make sure you've selected the correct answer choice and haven't made a costly careless mistake (such as marking an answer choice that you didn't mean to mark). This quick double check should more than pay for itself in caught mistakes for the time it costs.

⊘ PACE YOURSELF

It's easy to be overwhelmed when you're looking at a page full of questions; your mind is confused and full of random thoughts, and the clock is ticking down faster than you would like. Calm down and maintain the pace that you have set for yourself. Especially as you get down to the last few minutes of the test, don't let the small numbers on the clock make you panic. As long as you are on track by monitoring your pace, you are guaranteed to have time for each question.

⊘ DON'T RUSH

It is very easy to make errors when you are in a hurry. Maintaining a fast pace in answering questions is pointless if it makes you miss questions that you would have gotten right otherwise. Test writers like to include distracting information and wrong answers that seem right. Taking a little extra time to avoid careless mistakes can make all the difference in your test score. Find a pace that allows you to be confident in the answers that you select.

⊘ KEEP MOVING

Panicking will not help you pass the test, so do your best to stay calm and keep moving. Taking deep breaths and going through the answer elimination steps you practiced can help to break through a stress barrier and keep your pace.

Final Notes

The combination of a solid foundation of content knowledge and the confidence that comes from practicing your plan for applying that knowledge is the key to maximizing your performance on test day. As your foundation of content knowledge is built up and strengthened, you'll find that the strategies included in this chapter become more and more effective in helping you quickly sift through the distractions and traps of the test to isolate the correct answer.

Now that you're preparing to move forward into the test content chapters of this book, be sure to keep your goal in mind. As you read, think about how you will be able to apply this information on the test. If you've already seen sample questions for the test and you have an idea of the question format and style, try to come up with questions of your own that you can answer based on what you're reading. This will give you valuable practice applying your knowledge in the same ways you can expect to on test day.

Good luck and good studying!

Study Questions

NOTE FROM THE EDITOR:

Questions in this section are organized by topic, and answers follow each question. This is so that you can get a feel for what types of questions you do well on and which you may need to study more. The next section is laid out as a practice test, with randomized question order. It is recommended to study the questions and answers in this section, then test yourself on the second set of questions.

DEFINITIONS

1. A branch circuit that consists of two or more ungrounded conductors with a voltage between them and a neutral conductor with equal voltage between it and each ungrounded conductor of the circuit and that is connected to the neutral conductor of the system is best described as _____.

    a. Branch circuit, general-purpose
    b. Branch circuit, individual
    c. Branch circuit, multiwire
    d. Branch circuit, appliance

1. C: Article 100 contains general definitions intended to apply wherever they are used throughout the codebook. After consulting Article 100, the definition in the question is best described as branch circuit, multiwire.

2. The NEC defines *concealed* as _____.

    a. Rendered inaccessible by the structure or finish of the building
    b. Capable of being removed or exposed without damaging the building structure or finish
    c. Surrounded by a case, housing, fence, or wall(s) that prevents persons from accidentally contacting energized parts
    d. Kept secret; hidden

2. A: Article 100 Part 1 contains definitions intended to apply generally wherever they are used throughout the codebook. XXX.2 sections of other articles also contain definitions. After consulting Article 100, answer **a** is the correct definition of *concealed* as defined by the NEC.

CALCULATIONS

3. Determine the voltage drop of a single-phase 240-volt circuit, feeding a 1,000-watt metal-halide flood light that draws 4.17 amperes. The fixture is located 300 feet away from the panelboard and is wired with no. 12 AWG copper conductors.

    a. 4.94 volts
    b. 5.33 volts
    c. 9.89 volts
    d. 1.23 volts

3. A: Determining voltage drop within a circuit can be determined by using the following formula:

$$V_d = \frac{2(K)(I)(D)}{CM}$$

V_d = Voltage drop

K = Direct current constant, 12.9 ohms for copper conductors, 21.2 ohms for aluminum conductors.

I = Current, 4.17 amperes

D = One-way length of conductors to the load, 300 feet.

CM = Area in circular mils per Chapter 9 table 8, 6,530 circular mils.

Single-phase circuits will have two current-carrying conductors; therefore, the numerator must be multiplied by two to account for the current, length, and inherent resistance within the circuit to the load and back to the supply.

$$V_d = \frac{2(12.9)(4.17)(300)}{6,530}$$
$$V_d = 4.94 \text{ volts}$$

4. Ignoring slip, calculate the revolutions per minute (RPM) of a four-pole, three-phase induction motor operating at 35 hertz.
   a.   3600 RPM
   b.   1800 RPM
   c.   1050 RPM
   d.   980 RPM

4. C: To calculate the RPM of a three-phase induction motor, use the following formula:

$$RPM = \frac{120 \times frequency \text{ (Hz)}}{Number \ of \ poles}$$
$$RPM = \frac{120 \times 35 \text{ Hz}}{4}$$
$$RPM = \frac{4200}{4}$$
$$RPM = 1050$$

Unless a motor is controlled by a variable frequency drive, it will be operating at the frequency of the supplied power, either 60 Hz in the United States or 50 Hz for most European countries. Motor nameplates will list the speed at which the motor operates, whether at 60 Hz, 50 Hz, or both. When the motor controller is a VFD, the speed of the motor can be controlled by varying the frequency of the current to the motor.

5. Calculate the total resistance of ten 100-ohm resistors connected in parallel.

    a. 1 ohm
    b. 10 ohms
    c. 100 ohms
    d. 1,000 ohms

5. B: A general rule of thumb is that the total resistance of a group of parallel resistors will always be smaller than the smallest resistor. The following formula is capable of being used to solve for total resistance of any type of parallel resistor network.

$$R_T = \frac{1}{\frac{1}{R_1} + \frac{1}{R_2} + \frac{1}{R_3} \cdots}$$

Or:

$$\frac{1}{R_T} = \frac{1}{R_1} + \frac{1}{R_2} + \frac{1}{R_3} \cdots$$

$$\frac{1}{R_T} = \frac{1}{100} + \frac{1}{100} + \frac{1}{100} + \frac{1}{100} + \frac{1}{100} + \frac{1}{100} + \frac{1}{100} + \frac{1}{100} + \frac{1}{100} + \frac{1}{100}$$

$$R_T = \frac{1}{0.1}$$

$$R_T = 10 \text{ ohms}$$

THEORY

6. Which of the following is true when lowering alternating current frequency in a purely capacitive circuit?

    a. Lower total current flowing in the circuit
    b. Raise capacitive reactance
    c. Increase capacitance
    d. Both a and b

6. D: Capacitors in alternating-current circuits oppose current flow in the form of capacitive reactance. The capacitive reactance (X_c) provided by a capacitor is based primarily on two factors: the frequency (f) of the alternating current and the capacitance (C) of the capacitor. The unit for frequency is hertz (Hz) and the unit for capacitance is the farad (F).

$$X_c = \frac{1}{2\pi f C}$$

Lowering the frequency of an alternating-current circuit will result in an increase in capacitive reactance (X_c), thus increasing impedance and lowering total current flow.

PLANS

7. The following symbol found on a construction plan would most likely indicate what electrical device?

    a. A single receptacle
    b. A ceiling-mounted lighting outlet
    c. A wall-mounted lighting outlet
    d. Stairway lighting

7. C: It is not uncommon for a wall-mounted lighting outlet to be mistaken for a single or simplex receptacle. The single line attaching the circle to the wall on which it will be mounted is sometimes mistaken for the single line seen on the symbol for a single receptacle. The important distinction between the two is the lack of any lines within the circle for lighting outlets. A lighting outlet that has a line attaching it to a wall indicates that the lighting outlet will be wall-mounted. Ceiling-mounted lighting outlets will often be placed towards the center of a room and have no lines connecting to walls.

8. A blueprint scale reads *3/16" = 1'0"*. On the blueprint, a floor receptacle measures two and thirteen-sixteenths (2 ¹³/₁₆ inches) from a wall. What should the measurement be from the wall to the floor receptacle in the actual installation?

    a. 4 feet 4 inches
    b. 10 feet
    c. 15 feet
    d. 21 feet 3 inches

8. C: A *3/16" to 1'0"* scale is also known as a 1:64 scale. A measurement of *2 ¹³/₁₆"* on a blueprint will translate to a measurement of 15 feet at the actual installation.

$$2\frac{13}{16} = 2.8125 \quad \text{and} \quad \frac{3}{16} = 0.1875$$

$$\frac{2.8125}{0.1875} = 15 \text{ feet}$$

ELECTRICAL SERVICES

9. Service entrance conductors are considered to be outside of a building under all of the following conditions, EXCEPT _____.

    a. Where installed under no less than 2 inches of concrete beneath a building or other structure.

    b. Where installed within a building or other structure in a raceway encased in concrete or brick no less than 2 inches thick.

    c. Where installed in conduit and under no less than 18 inches of earth beneath a building or other structure.

    d. Where installed in rigid metal conduit (type RMC) or intermediate metal conduit (type IMC) in the interstitial space between floors of a building or other structure.

9. D: Article 230.6 lists all the conditions for which service entrance conductors are considered to be *outside*. This is important due to the requirements of 230.70(A)(1) requiring interiorly located disconnecting means to be located at the nearest point of entrance. Where service entrance conductors are contained within 2 inches of brick or concrete, even in vertical runs within the footprint of the building, they are still considered to be outside. The code section makes no such exceptions for interstitial spaces.

10. Any circuit conductors other than service entrance conductors shall not be installed in the same raceway, cable, handhole enclosure, or underground box as the service conductors are installed, unless meeting which of the following requirements?

    a. The conductor(s) is/are load management control conductors having overcurrent protection.

    b. The conductor(s) is/are of the same size and insulation type as the service entrance conductors.

    c. The raceway in which the conductors are contained is sized at least 250% larger than would be required for the combination of conductors.

    d. Conductors other than service entrance conductors are never allowed to be installed in the same raceway in which service conductors are installed.

10. A: 230.7 states that conductors other than service conductors shall not be installed in the same raceway, cable, handhole enclosure, or underground box as the service conductors. There are two exceptions: grounding electrode conductors or supply-side bonding jumpers, and load management conductors having overcurrent protection. Noteworthy is the fact that *gutters* do not meet the definition of a *raceway* per Article 100, and therefore are allowed to house both service entrance conductors and other conductors.

11. Service conductors installed as open conductors or multiconductor cables without an overall outer jacket shall have a clearance of no less than ____ from openable windows, doors, porches, balconies, ladders, stairs, fire escapes, or similar locations.

    a. 1 foot
    b. 2 feet
    c. 3 feet
    d. 4 feet

11. C: 230.9(A) states that 3 feet is the required clearance away from windows, doors, and similar locations. The 3-foot clearance does not apply to service entrance conductors contained within a raceway or having an overall outer jacket, such as type SE cable. The exception allows for conductors above the top level of a window to be less than 3 feet.

12. Except in the case of conductors supplying only limited loads of a single branch circuit, overhead service conductors shall not be sized smaller than ____.

    a. 1/0 AWG copper or 2/0 AWG aluminum
    b. 1 AWG copper or 1/0 AWG aluminum
    c. 4 AWG copper or 3 AWG aluminum
    d. 8 AWG copper or 6 AWG aluminum

12. D: Per 230.23(B), the minimum size overhead service conductor is 8 AWG copper or 6 AWG aluminum, with the exception of conductors that supply limited loads of a single branch circuit. These conductors shall not be smaller than 12 AWG hard-drawn copper or equivalent.

SERVICE EQUIPMENT

13. A surge-protective device shall be provided with all services supplying ____.

    a. Healthcare facilities
    b. Commercial buildings
    c. Sensitive electronic equipment
    d. Dwelling units

13. D: The NEC includes a requirement for services supplying dwelling units, dormitory units, guest room of hotels and motels, as well as areas of nursing homes and limited care facilities used exclusively as patient sleeping rooms. 230.67(A) states that all services supplying these occupancies must be provided with a surge-protective device (SPD). Also important is 230.67(D), which requires that an SPD be provided when replacing existing service equipment.

14. When meeting the requirements within Article 230 Part VI, what is the maximum number of disconnecting means permitted for each service or set of service-entrance conductors?

    a. 1
    b. 2
    c. 4
    d. 6

14. D: Each set of service-entrance conductors is permitted to supply up to six service disconnects in lieu of a single main disconnect, provided they meet the requirements of 230.71(B). The NEC has previously referred to this as *six operations of the hand*.

15. Each service disconnect shall simultaneously disconnect all _____ conductors that it controls from the premises wiring system.

    a. Ungrounded
    b. Grounded
    c. Grounding
    d. Bonding

15. A: In general, service disconnects are required to simultaneously disconnect all ungrounded conductors per 230.74. Service disconnecting means sometimes disconnect the grounded conductor, but it is not generally a requirement.

SERVICE LOAD CALCULATIONS

16. Using the optional method, calculate the size of the ungrounded conductors required to feed a one-family dwelling unit whose floor area is 2,300 square feet, not including an unfinished cellar, unfinished attic, and open porch. This will have a 10-kW Range, a 4.5-kW water heater, a 1.2-kW dishwasher, and a 4.5-kW clothes dryer. The dwelling unit is climate controlled with 15-kW of electric heat and a 2-ton (30 ampere @ 230 volt) central air conditioner.

    a. 3 AWG copper
    b. 2 AWG copper
    c. 1 AWG copper
    d. 1/0 AWG copper

16. B: The requirements for determining the service load of dwelling units using the optional method can be found in Part IV of Article 220. Care should be taken to ensure the correct calculated load is obtained. Article 220 can be somewhat complicated for those unfamiliar with the intent of the optional method. However, the steps necessary to calculate the load correctly can be distilled into just a few key ideas, many of them applying to all such calculations. For this article, it is generally accepted that kilovolt-amperes (kVA) shall be considered equivalent to kilowatts (kW).

The first step is to determine the demand from the general loads. 220.82(B) lists several loads that will be part of every service calculated with the optional method.

220.82(B)(1): 3 volt-amperes per square foot.

220.82(B)(2): All dwelling units are required to have two 20-ampere small-appliance branch circuits at 1,500 volt-amperes each, and each laundry branch circuit.

220.82(B)(3): The nameplate rating of appliances that are permanently connected or fastened in place. This includes ranges, wall-mounted ovens, cooktops, clothes dryers, water heaters, etc.

220.82(B)(4): The nameplate ampere or kVA rating of all permanently connected motors not included in item (3).

100% of the sum of the items above applies to the first 10kVA, and 40% of the remaining load.

Heating and air conditioning loads are determined by 220.82(C). Electricians should use one of the following six items in addition to the general loads calculation.

220.82(C)(1): 100% of the nameplate rating of the air conditioning and cooling.

220.82(C)(2): 100% of the nameplate rating of the heat pump when used without supplemental electric heating.

220.82(C)(3): 100% of the nameplate rating of the heat pump compressor and 65% of the supplemental heat, unless the two are prevented from operating at the same time. Then only 65% of the supplemental heat is used.

220.82(C)(4): 65% of the nameplate rating of all electric space heating if less than four separately controlled units. Homes with central heating having less than four separate units will fall into this category.

220.82(C)(5): 40% of the nameplate rating of all electric space heating if more than four separately controlled units. This is more common with heating units installed in individual rooms within a dwelling unit.

220.82(C)(6): 100% of the nameplate rating of electric thermal storage and other heating systems where the usual load is expected to be continuous.

The sum of both the demand factor-adjusted general loads and the heating and air conditioning loads is the optional calculated load of a feeder or service. Therefore, the calculation of the load on the service in question can be obtained as illustrated below.

General Load

General lighting and receptacles at 3 VA per square foot: 6,900 VA
Two small appliance branch circuits at 1,500 VA each: 3,000 VA
Laundry circuit: 1,500 VA
Range: 10,000 VA
Water heater: 4,500 VA
Dishwasher: 1,200 VA
Dryer: 4,500 VA
Total general load: 31,600 VA

Application of Demand Factor

100% of first 10 kVA: 10,000 VA
40% of remaining load: 8,640 VA
Subtotal of general load: 18,640 VA

Heating and Air Conditioning Loads

Heat at 65% of nameplate rating (15 kW): 9,750 VA
Air conditioning at 100% of nameplate rating 230 V × 30 A = 6,900 VA

Total Calculated Load of Service

Subtotal of general load: 18,640 VA
Largest of heat and air conditioning (heat): 9,750 VA
Total calculated load: 28,390 VA

Determining Ungrounded Conductor Sizing

$$\frac{28{,}390 \text{ VA}}{240 \text{ V}} = 118 \text{ A}$$

The calculated load is 118 amperes. At a minimum, a calculated load of 118 amperes must be provided. Because this does not correspond to a standard ampere rating, the next highest rating found in 240.6(A) must be used. Thus, the service size is 125 amperes.

When sizing ungrounded service-entrance conductors for single-phase dwelling units, 310.12 stipulates that, for services rated 100 amperes through 400 amperes, the service conductors are permitted to have an ampacity no less than 83% of the service rating.

$$125 \text{ A} \times 0.83 = 104 \text{ A}$$

Additionally, when no other adjustment factors or corrections are required, Table 310.12 is permitted to be used in determining the conductor size for a given service rating. Either by using Table 310.12 or by using Table 310.16 and 75 °C column, the correct ungrounded conductor size for this service is **2 AWG copper**.

17. Using the optional method, determine the proper service size for a one-family dwelling unit whose square footage is 3,500 square feet. Within this, 250 square feet of the total consists of an unfinished cellar not adaptable for future use. The owner has requested two dedicated laundry circuits for cord- and plug-connected loads. Appliances connected to the service consist of two 6-kW wall-mounted double ovens, two 5-kW water heaters, a 5-kW clothes dryer, a 1,000 VA dishwasher, and a 1,000 VA disposal. The air conditioning is a 5-ton unit (31 amperes at 230 volts), and 10-kW of heating with five independently controlled room heaters.

    a. 200 amperes
    b. 175 amperes
    c. 150 amperes
    d. 125 amperes

17. C: The requirements for determining the service load of dwelling units using the optional method can be found in Part IV of Article 220. Care should be taken to ensure the correct calculated load is obtained. Article 220 can be somewhat complicated for those unfamiliar with the intent of the optional method. However, the steps necessary to calculate the load correctly can be distilled into just a few key ideas, many of them applying to all such calculations. For this article, it is generally accepted that kilovolt-amperes (kVA) shall be considered equivalent to kilowatts (kW).

The first step is to determine the demand from the general loads. 220.82(B) lists several loads that will be part of every service calculated with the optional method.

220.82(B)(1): 3 volt-amperes per square foot. This does not include garages, porches, or other unfinished spaces.
220.82(B)(2): All dwelling units are required to have two 20-ampere small-appliance branch circuits at 1,500 volt-amperes each, and each laundry branch circuit.
220.82(B)(3): The nameplate rating of appliances that are permanently connected or fastened in place. This includes ranges, wall-mounted ovens, cooktops, clothes dryers, water heaters, etc.
220.82(B)(4): The nameplate ampere or kVA rating of all permanently connected motors not included in item (3).
100% of the sum of the items above applies to the first 10 kVA, and 40% of the remaining load.

Heating and air conditioning loads are determined by 220.82(C). One of the following six items should be used, in addition to the general loads calculation.

220.82(C)(1): 100% of the nameplate rating of the air conditioning and cooling.
220.82(C)(2): 100% of the nameplate rating of the heat pump when used without supplemental electric heating.
220.82(C)(3): 100% of the nameplate rating of the heat pump compressor and 65% of the supplemental heat, unless the two are prevented from operating at the same time. Then only 65% of the supplemental heat is used.
220.82(C)(4): 65% of the nameplate rating of all electric space heating if less than four separately controlled units. Homes with central heating having less than four separate units will fall into this category.
220.82(C)(5): 40% of the nameplate rating of all electric space heating if more than four separately controlled units. This is more common with heating units installed in individual rooms within a dwelling unit.
220.82(C)(6): 100% of the nameplate rating of electric thermal storage and other heating systems where the usual load is expected to be continuous.

The sum of both the demand factor adjusted general loads and the heating and air conditioning loads is the optional calculated load of a feeder or service. Therefore, the calculation of the required service size in question can be obtained as illustrated below.

General Load

General lighting and receptacles at 3 VA/square foot (3,500 square feet – 250 square feet × 3 VA): 9,750 VA
Two small appliance branch circuits at 1,500 VA each: 3,000 VA
Two laundry circuit: 3,000 VA
Range: 12,000 VA
Two water heaters: 10,000 VA

Dryer: 5,000 VA
Dishwasher: 1,000 VA
Disposal: 1,000 VA
Total general load: 44,750 VA

Application of Demand Factor

100% of first 10 kVA: 10,000 VA
40% of remaining load: 13,900 VA
Subtotal of general load: 23,900 VA

Heating and Air Conditioning Loads

Heat at 40% of nameplate rating (five separately controlled units): 4,000 VA
Air conditioning at 100% of nameplate rating (230 V × 31 A): 7,130 VA

Total Calculated Load of Service

Subtotal of general load: 23,900 VA
Largest of heat and air conditioning (A/C): 7,130 VA
Total calculated load: 31,030 VA

Determining Service Size

$$\frac{31,030 \text{ VA}}{240 \text{ V}} = 129 \text{ A}$$

The calculated load in amperes is 129 amperes. This means that, at a minimum, the calculated load of 129 amperes must be provided. Because this does not correspond to a standard ampere rating, the next highest rating found in 240.6(A) must be used. Thus, the service size is **150 amperes**.

SEPARATELY DERIVED SYSTEMS

18. Which of the following is an example of a separately derived system as defined by the NEC?

    a. A sub-panel fed from a thermal-magnetic circuit breaker on the load side of the service disconnect.
    b. A three-phase, delta primary, wye secondary transformer whose grounded conductor is derived from the common connection of three secondary phase windings.
    c. A generator sharing a common, unswitched grounded conductor with the primary service on a building.
    d. An uninterruptible power source (UPS) having a maintenance bypass that disconnects three ungrounded conductors and an unswitched grounded conductor.

18. B: Article 100 defines a separately derived system as an electrical power supply outlet other than a service having no direct connection(s) to circuit conductors of any source other than those established by grounding and bonding connections. Generators, UPS systems, photovoltaic systems, and transformers are all common examples of separately derived systems, provided that all circuit conductors are switched or otherwise not directly connected to another source.

19. Which of the following is not an acceptable means of grounding a separately derived alternating-current system?

    a. Installation of a system bonding jumper at both the source and the first means of disconnect
    b. Installation of a system bonding jumper at the source only
    c. Installation of a system bonding jumper at the first means of disconnect only
    d. Both b and c

19. D: 250.30(A)(1) requires that an unspliced system bonding jumper be made at any single point on a separately derived system from the source to the first system disconnecting means or overcurrent protection. In practice, this means that electricians should install the system bonding jumper in either the enclosure of the separately derived system, i.e., bonding the XO terminal, GEC, and grounded conductor to the same point in a transformer enclosure; or installing a screw-type bonding jumper in the panel supplied by the transformer. Doing both would provide a parallel current flow path for neutral current to flow. Often this leads to objectionable current flowing on normally non-current carrying parts of a system or premises.

ELECTRICAL FEEDERS

20. Where a feeder supplies continuous loads or any combination of continuous and noncontinuous loads, the minimum feeder conductor size shall have an ampacity no less than the noncontinuous loads plus _____ percent of the continuous load.

    a. 100
    b. 115
    c. 125
    d. 150

20. C: As is common throughout the NEC, conductors supplying a combination of continuous and noncontinuous loads must have an ampacity of 100% of the noncontinuous loads and 125% of the continuous loads[1]. 215.2(A)(1) stipulates that the larger of either the combination of this calculation, [215.2(A)(1)(a)] or be based off of the maximum load to be served after any adjustment or correction factors [215.2(A)(1)(b)].

1: The NEC defines a continuous load as one where the maximum current is expected to continue for three hours or more.

21. Conductors for feeders, as defined in Article 100, sized to prevent a voltage drop exceeding _____ at the farthest outlet of power, heating, lighting, or combinations of such loads, and where the maximum total voltage drop does not exceed _____, will provide reasonable efficiency of operation.

    a. 1%, 3%
    b. 3%, 5%
    c. 5%, 8%
    d. 5%, 10%

21. B: 215.2(A)(1) Informational note no. 2. Informational notes, while not mandatory, nevertheless provide important considerations when attempting to fully understand the intent of a code section. The NEC recommends that voltage drop on conductors for feeders not exceed 3% and that the total voltage drop on both feeders and branch circuits not exceed 5%. Conversely, when considering branch circuits alone, the NEC recommends that voltage drop not exceed 3%, and the total of both branch circuits and feeders not exceed 5%.

22. Where the assembly, including the overcurrent devices protecting the feeder(s), is listed for operation at 100% of its rating, the overcurrent device shall be permitted to be not less than 100% of the noncontinuous load plus ____ percent of the continuous load.

    a. 100
    b. 115
    c. 124
    d. 80

22. A: 215.3 Exception. While it is generally required to size overcurrent protection to 100% of the noncontinuous load plus 125% of the continuous load, 215.3 Exception states that when the assembly, including the overcurrent protection device protecting feeders, is listed for operation at 100% of its rating, the overcurrent device shall be permitted to be not less than the sum of the continuous load plus the noncontinuous load.

BRANCH CIRCUIT CALCULATIONS AND CONDUCTORS

23. A multiwire branch circuit shall be permitted to be considered as multiple circuits. All conductors of a multiwire branch circuit shall ____.

    a. Originate from the equipment containing the branch-circuit overcurrent protective device(s)
    b. Be listed for use in a multiwire branch circuit
    c. Bear a continuous outer stripe of the same color, other than white, gray, or green, along the entire length of the conductor
    d. Terminate on the same device yoke intended for use by only one utilization equipment

23. A: 210.4(A). A multiwire branch circuit—defined by the NEC as a branch circuit that consists of two or more ungrounded conductors that have a voltage between them and a neutral conductor that has equal voltage between it and each ungrounded conductor of the circuit and that is connected to the neutral conductor of the system—is permitted to be considered as multiple circuits. All conductors must originate from the equipment containing the branch-circuit overcurrent protective device(s), considering a multiwire branch circuit to multiple circuits is often used to allow a single cable to meet the requirements of providing two small-appliance branch circuits for countertop receptacles in dwelling unit kitchens.

Copyright © Mometrix Media. You have been licensed one copy of this document for personal use only. Any other reproduction or redistribution is strictly prohibited. All rights reserved. This content is provided for test preparation purposes only and does not imply an endorsement by Mometrix of any particular political, scientific, or religious point of view.

24. Each multiwire branch circuit shall be provided with a(n) _____.

    a. Permanent means of identifying the neutral conductor as a part of a multiwire branch circuit at all splice points
    b. Identifying marking on either the dead front cover or panel schedule indicating the OCPD(s) supplying the multiwire branch circuit
    c. Means to simultaneously disconnect all ungrounded conductors at the point where the branch circuit originates
    d. Ground-fault circuit interruption protection for personnel

24. C: 210.4(B). Multiwire branch circuits are commonly found in existing applications. As of the 2008 NEC, all multiwire branch circuits are required to be provided with a means to simultaneously disconnect all ungrounded conductors. This is to prevent maintenance personnel from unknowingly coming in contact with a grounded (neutral) conductor that is still carrying a load from the remaining conductors. In the event the continuity of the grounded conductor is interrupted, a difference in potential between the section of the grounded conductor from the load and the section of grounded conductor still referenced to ground within the distribution panel would develop.

25. The ungrounded and grounded conductors of each multiwire branch circuit shall be grouped by any of the following means, EXCEPT:

    a. Wire markers.
    b. Cable ties.
    c. Originate from a cable or raceway unique to the circuit that makes the grouping obvious.
    d. Routed down to the associated OCPD and then up to the point at which it is connected to the grounded conductor within the enclosure.

25. D: 200.4(B), 210.4(D). When an enclosure or panel contains multiple circuits where more than one neutral conductor is associated with different circuits, the ungrounded conductors shall be grouped with the associated neutral conductors in at least one place within the enclosure. This can be achieved by either matching wire markers, cable ties, or similar means. If a set of conductors originate from the same cable or raceway, grouping in this fashion is not required. Additionally, if the conductors pass through a box or conduit body without a loop as described in 314.16(B)(1), or without a splice or termination, grouping is not required.

26. In dwelling units and guest rooms or guest suites of hotels, motels, and similar occupancies, the voltage shall not exceed _____ nominal between conductors that supply the terminals of luminaires or cord-and-plug connected loads, 1,440 volt-amperes nominal or less.

    a. 120 volts
    b. 150 volts
    c. 240 volts
    d. 250 volts

26. A: 210.6(A). In dwelling units and similar occupancies, luminaires and cord-and-plug connected loads shall not exceed 120 volts to ground. This is to limit occupants' possible exposure to shock hazards associated with higher voltage equipment found in other types of installations.

27. Circuits exceeding 120 volts nominal between conductors and not exceeding 277 volts nominal to ground shall be permitted to supply cord-and-plug connected or permanently connected utilization equipment or any of the following types of listed luminaires, EXCEPT:

    a. Electric-discharge luminaires with integral ballasts.
    b. Incandescent or LED luminaires, equipped with medium-base or smaller screw shell lampholders, where the lampholder is not supplied from the output of a stepdown autotransformer or driver integral to the luminaire.
    c. Luminaires equipped with mogul-base screw shell lampholders.
    d. Luminaires without lampholders.

27. B: 210.6(C). Where luminaires are supplied from circuits exceeding 120 volts-to-ground and not exceeding 277 volts-to-ground, medium base and smaller screw shell lampholders are not allowed unless the lampholder is supplied at 120 volts or less from the output of a stepdown transformer, LED driver, or similar integral component of the luminaire.

28. Which of the following locations within a dwelling unit does NOT require the installations of ground-fault protection for personnel?

    a. Receptacles serving kitchens
    b. Receptacles located less than 6 feet from the top inside edge of the bowl of a sink or outside edge of a bathtub or shower within a bathroom
    c. Receptacles located more than 6 feet from the top inside edge of the bowl of a sink or outside edge of a bathtub or shower not in a bathroom
    d. Areas with sinks and permanent provisions for food

28. C: 210.8(A). Though not a common installation, receptacles more than 6 feet from the edge of sinks, bathtubs, or showers not located in bathrooms or other locations specified in 210.8(A) are not required to have GFCI protection. If a room meets the NEC definition of bathroom, all receptacles within are required to have GFCI protection, regardless of their distance from sinks, etc.

29. Which of the following dwelling unit lighting outlets require ground-fault protection of personnel?

    a. Lighting outlets underneath covered porches
    b. Lighting outlets not exceeding 120 volts installed in crawl spaces
    c. Lighting in bathrooms containing a steam shower
    d. Boathouse lighting outlets

29. B: 210.8(C). GFCI protection is required for lighting outlets installed in crawl spaces. This is due to reduced space between the ground and the light fixture. GFCI protection is not required for any other lighting outlets.

30. All dwelling units are required to be provided with ___ or more 20-ampere small-appliance branch circuits and, excluding exceptions, shall have no other outlets.

    a. 1
    b. 2
    c. 3
    d. No minimum requirement

30. B: 210.11(C)(1) requires, in addition to other parts of 210.11, two or more 20-ampere small-appliance branch circuits. Excluding the exception for the sole connection of an electric clock in any of the rooms specified in 210.52(B)(1), and the exception allowing the addition of a receptacle to supply the power for supplemental equipment and lighting on gas-fired ranges, ovens, or counter-mounted cooking units, these circuits shall have no other outlets.

31. What is the maximum number of bathrooms that can be supplied by a single 120-volt, 20-ampere branch circuit to supply bathroom receptacle outlets?

    a. 1
    b. 2
    c. 3
    d. No maximum number

31. D: 210.11(C)(3). The NEC does not specify a maximum number of bathrooms that can be supplied by a single 120-volt, 20-ampere branch circuit. When supplying more than one bathroom, the bathroom branch circuit shall have no other outlets.

32. Where a single 120-volt, 20-ampere branch circuit supplies only one bathroom, which of the following is permitted to be supplied for the same circuit?

    a. A vanity light fixture and exhaust fan with a maximum load of 8 amperes
    b. Four recessed light fixtures and a radiant ceiling heater with a maximum load of 11.5 amperes
    c. Outdoor receptacles required to be within 25 feet of equipment likely to require service
    d. No additional outlets are allowed for this type of circuit.

32. A: 210.11(C)(3) Exception and 210.23(A)(1) and (A)(2). Where a bathroom branch circuit supplies only a single bathroom, it is permitted to serve other outlets or equipment within the same bathroom, provided that the combined rating of utilization equipment does not exceed 50% of the branch-circuit ampere rating (10 amperes).

33. When required to be installed in dwelling units, dormitory units, or other occupancies, arc-fault circuit interrupters (AFCI) shall be listed and installed in a(n) _____ location.

    a. Accessible
    b. Readily accessible
    c. Exposed
    d. Concealed

33. B: 210.12. Arc-fault circuit interrupters (AFCI) are required to be installed in a readily accessible location. This is due not only to the fact that they may require resetting, but also because AFCIs should be tested monthly or as recommended by the manufacturer. Note that the NEC differentiates between accessible and readily accessible locations, wherein the latter must be able to be reached quickly for routine operation or inspection.

34. Where a combination-type AFCI circuit breaker is installed at the origin of the branch circuit, what additional installation requirements are necessary?
    a. The first outlet shall be marked to indicate that it is the first outlet on the circuit.
    b. The maximum length of the branch-circuit wiring from the overcurrent device to the first outlet is 50 feet for 14 AWG and 70 feet for 12 AWG conductors.
    c. A metal raceway, metal wireway, metal auxiliary gutter, or type MC or AC cable are installed for the portion of the branch circuit between the branch-circuit overcurrent device and first outlet.
    d. No additional installations requirements are necessary.

34. D: 210.12(A)(1). When a listed combination-type arc-fault circuit interrupter is installed to provide protection to the entire branch circuit, no additional installation requirements are necessary. Other types of AFCI protection, such as listed outlet branch-circuit type AFCIs installed at the first outlet, will require additional measures to ensure the protection of the entire circuit.

35. Determine the minimum size copper conductor required to supply a branch circuit with a combination of 10 amperes of continuous load and 10 amperes of non-continuous load.
    a. 14 AWG
    b. 12 AWG
    c. 10 AWG
    d. 8 AWG

35. C: 210.19(A)(1) requires the ampacity of a branch circuit to be based on 125% of the continuous load plus 100% of the non-continuous load.

Continuous load = 10 amperes × 1.25 = 12.5 amperes

Non-continuous load = 10 amperes

Total load = 22.5 amperes

With a calculated load of 22.5 amperes, the next highest standard ampere rating (Table 240.6) of 25 amperes is required. Table 310.16 is used to determine the ampacity of insulated conductors with no more than three current-carrying conductors in a raceway, cable, or directly buried. It is common practice to use the 75 °C column due to the temperature limitations of nearly all terminations and outlets. We find 12 AWG with a listed ampacity of 25 amperes. However, referencing the notes, section 240.4(D) requires that overcurrent protection for 12 AWG copper conductors not exceed 20 amperes. Thus, electricians must move to the next size copper conductor, 10 AWG.

36. What is the allowable current rating(s) for receptacles installed on a branch circuit rated 20 amperes supplying two or more receptacles for the connection of cord-and-plug connected loads?

    a. 15 amperes
    b. Either 15 or 20 amperes
    c. 20 amperes
    d. 30 amperes

36. B: When receptacles are installed on circuits supplying more than one receptacle, Table 210.21(B)(3) is used to determine the rating of the installed receptacles. Electricians are required to install only receptacles rated 15 amperes on branch circuits rated 15 amperes. However, they are allowed to install either 15-ampere or 20-ampere receptacles on branch circuits rated 20 amperes. Though somewhat counterintuitive, a proper understanding of the purpose of this code section makes clear the importance of its requirement.

The installation of a 20-ampere receptacle allows for the connection of devices with a 20-ampere male cord plug. Such devices are provided with a horizontal grounded (neutral) prong, which serves as a means to prevent the connection of 20-ampere devices to a 15-ampere receptacle. Additionally, it is important to understand that 15-ampere receptacles are rated for 20 amperes of feed-through current and thus allowed to be installed on 20-ampere circuits.

37. What is the allowable current rating(s) for a single receptacle installed on an individual branch circuit rated 20 amperes for the connection of cord-and-plug connected loads?

    a. 15 amperes
    b. Either 15 or 20 amperes
    c. 20 amperes
    d. 30 amperes

37. C: Though 210.21(B)(3) allows for the installation of receptacles rated both 15 and 20 amperes to a branch circuit rated for 20 amperes, 210.21(B)(1) requires that single receptacles installed on an individual branch circuit be rated no less than the rating of the circuit. Of note is the fact that, by definition, a duplex receptacle is not considered a single receptacle, but multiple receptacles. This code section refers to receptacles containing only a single receptacle on the device yoke.

38. For 15- and 20-ampere branch circuits supplying two or more outlets or receptacles, the rating of any one cord-and plug-connected utilization equipment not fastened in place shall not exceed _____ of the branch-circuit ampere rating.

    a. 50%
    b. 75%
    c. 80%
    d. 100%

38. C: 210.23(B)(1) states that, for 15- and 20-ampere branch circuits, the rating of any one cord-and-plug connected utilization equipment not fastened in place shall not exceed 80% of the branch-circuit ampere rating.

39. For 15- and 20-ampere branch circuits supplying lighting units or other outlets, as well as utilization equipment fastened in place, the total rating of the utilization equipment fastened in place, other than luminaires, shall not exceed _____ of the branch-circuit ampere rating.

    a. 50%
    b. 75%
    c. 80%
    d. 100%

39. A: 210.23(B)(2). When utilization equipment is fastened in place and installed on the same 15- or 20-ampere branch circuit as lighting or other outlets, the total rating of the utilization equipment cannot exceed 50% of the branch-circuit ampere rating. This requirement does not apply when the branch circuit supplies only equipment fastened in place, such as a branch circuit supplying both a dishwasher and a microwave.

40. Determine the number of countertop receptacles required to be installed in the kitchen shown in the figure below.

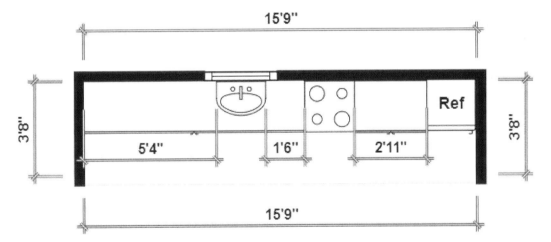

    a. 4
    b. 5
    c. 6
    d. 7

40. A: When installing countertop receptacles, the installer must ensure that a device can be placed anywhere on the counter and be within 2 feet of a receptacle. At least one receptacle is required to be installed on countertop spaces in 12 inches or wider in kitchens and other similar areas of a dwelling unit. Receptacle outlets shall not be required directly behind a range, counter-mounted cooking unit, or sink. The countertop section from the left requires two receptacles: one within 2 feet of the wall, and at least one more between the first receptacle and the edge of the sink. The countertop section between the sink and the cooktop and between the cooktop and the refrigerator are both larger than 1 foot and therefore require at least one receptacle each.

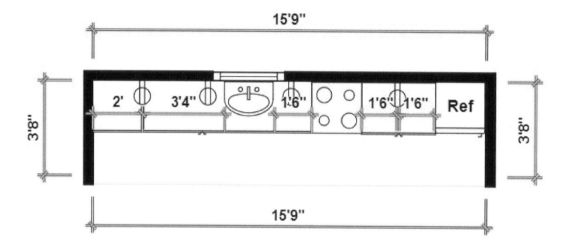

41. Determine the number of countertop receptacles required to be installed in the kitchen shown in the figure below. The island has a surface area of 8 square feet.

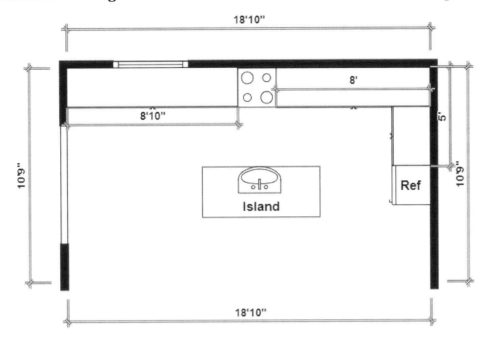

a. 5
b. 6
c. 7
d. 8

41. D: Starting at the upper-left countertop section, once receptacles are placed within 2 feet of both the wall and the range, there is 4' 10" of counter space between receptacles. A third receptacle is required. It is less clear what is appropriate for the countertop section between the range and the refrigerator. The language of the code section states that measurements must be taken horizontally. This is generally understood to mean horizontally along the wall and around corners, not cutting across a countertop. This makes more sense when considering that the boxes for these receptacles will be installed and inspected prior to the cabinets or countertops being installed. Thus when placing a receptacle within 2 feet of the range and within 2 feet of the refrigerator there is a 9-foot section of countertop left. Placing an additional receptacle 4 feet from one of the others leaves an area that would need a receptacle if it were not measured around a corner. The corner, however, appears to "bunch" the receptacles up once the countertop is installed. Nevertheless, the general interpretation of the code requires that an additional receptacle be installed in the corner. Lastly, any island with an area of nine square feet or less will require only one receptacle.

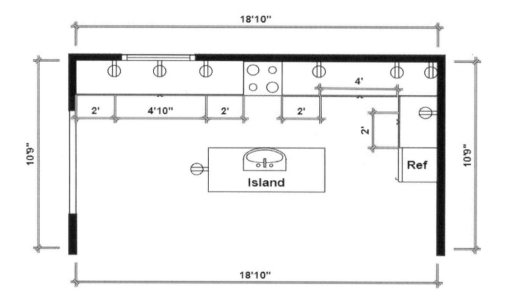

ELECTRICAL WIRING METHODS AND ELECTRICAL MATERIAL

42. In both exposed and concealed locations, where a cable- or raceway-type wiring method is installed through bored holes in joists, rafters, or wood members, holes shall be bored so that the edge of the hole is not less than _____ from the edges of the wood member.

    a. 1 inch
    b. 1¼ inches
    c. 1¾ inches
    d. 2 inches

42. B: 300.4(A)(1). When cable- or raceway-type wiring methods are installed through bored holes in framing members, they are required to be not less than 1¼ inches from the nearest edge of the framing member. For a standard 2 x 4 nominal (1 x ½ x 3½ actual) wall stud, this requires that a ¾-inch hole be bored within ¼-inch of the center of the stud.

43. Where the proper distance cannot be maintained, a cable or raceway shall be protected from penetration by nails and screws by _____ installed to cover the area of the wiring.

    a. A steel plate(s) or bushing(s) at least 1/16-inch thick, and of appropriate length and width
    b. Rigid metal conduit, intermediate metal conduit, rigid PVC conduit, RTRC, or electrical metallic tubing
    c. A listed and marked steel plate less than 1/16-inch thick that provides equal or better protection
    d. Any of the above

43. D: 300.4(A)(1) and 300.4(A)(2). When a 1¼-inch distance from the nearest edge of the framing member cannot be maintained, the cable must be protected by a 1/16-inch steel plate or a steel plate listed and marked as providing equivalent protection, or installed in RMC, IMC, rigid PVC, or EMT. Though these methods allow for the protection of the conductors, one must also ensure that the strength of the framing member is not compromised.

44. In both concealed and exposed locations, where a cable- or raceway-type wiring method is installed parallel to framing members, such as joists, rafters, or studs; or is installed parallel to furring strips, the cable or raceway shall be installed and supported so that the nearest outside surface of the cable or raceway is no less than _____ from the nearest edge of the framing member or furring strip.

    a. 1 inch
    b. 1¼ inch
    c. 1¾ inch
    d. 2 inches

44. B: 300.4(D). For cables- or raceway-type wiring methods installed parallel to framing members, they must be installed 1¼ inches from the nearest edge of the framing member. For cables installed parallel to furring strips, cables must be securely fastened so that a distance of 1¼ inches is maintained from the furring strips to prevent accidental damage from nails and screws.

45. Where raceways containing insulated circuit conductors larger than _____ enter a cabinet, a box, an enclosure, or a raceway, the conductors shall be protected by an identified fitting providing a smoothly rounded insulating surface, a listed metal fitting that has smoothly rounded edges, separation from the fitting or raceway using an identified insulating material securely fastened in place, or threaded hubs or bosses that are integral to the cabinet or box.

    a. 6 AWG
    b. 4 AWG
    c. 3 AWG
    d. 2 AWG

45. A: 300.4(G). State code tests often change the wording of a question such that it resembles the relevant code section but requires a thorough understanding of the section to answer the question correctly. Such is the case here. The wording of 300.4(G) states that insulated conductors 4 AWG or larger require the installation of one of the protection methods listed in the question. However, the wording of this question is changed to "conductors larger than." It stands to reason that "conductors larger than 6 AWG" can be substituted for "conductors 4 AWG and larger" without changing the intent of the code section.

46. The interior of enclosures or raceways installed underground shall be considered to be a _____.

    a. Dry location
    b. Damp location
    c. Wet location
    d. Underground location

46. C: 300.5(B). It should always be assumed, both because it is required by the NEC and because it is practically impossible to prevent the intrusion of moisture into a raceway installed underground, that such types of installation are considered "wet locations." This requires, at a minimum, that all wiring methods and conductors be listed for use in wet locations.

47. Determine the minimum cover requirements for a direct burial cable containing three 14 AWG conductors installed in a trench across a residential yard 25 feet from the nearest edge of the dwelling unit.

    a. 6 inches
    b. 12 inches
    c. 18 inches
    d. 24 inches

47. D: Table 300.5. Direct burial cables installed in a trench and not meeting the requirements of any of the locations listed on the left side of the chart would require the use of "Column 1 / All locations not specified below," or 24 inches. Much of the additional information mentioned in the question is irrelevant to the application of Table 300.5 and therefore should be ignored.

48. Determine the minimum cover requirements for eight type THHN conductors in a Schedule 40 PVC conduit installed under a building.

    a. 0 inches
    b. 12 inches
    c. 18 inches
    d. 24 inches

48. A: Table 300.5. When installed under a building, the minimum cover requirements are 0 inches, provided the cables or conductors are installed in a raceway or are type MC or type MI cable identified for direct burial.

49. Determine the minimum cover requirements for a 20-ampere, 120-volt residential branch circuit having GFCI protection installed under a residential driveway.

    a. 6 inches
    b. 12 inches
    c. 18 inches
    d. 24 inches

49. B: Table 300.5. For residential branch circuits 120 volts or less rated not more than 20 amperes and having GFCI protection, a minimum burial depth of 12 inches is required, both underneath residential driveways and all locations not specified in other rows of Table 300.5. Note 4 for Table 300.4 states that, where one of the wiring methods listed in column 1 through 3 is used for one of the circuit types in columns 4 and 5, the shallowest depth shall be permitted, allowing for the installation of direct burial cable at a depth of 12 inches as opposed to 24 inches if installed in the same manner as column 1.

50. In multiwire branch circuits, the continuity of a _____ shall not depend on device connections such as lampholders or receptacles, where the removal of such devices would interrupt the continuity.

    a. Grounded conductor
    b. Grounding conductor
    c. Equipment grounding conductor
    d. Ungrounded conductor

50. A: 300.13(B). Grounded conductors (neutrals) of multiwire branch circuits that supply receptacles, lampholders, or other such devices must not depend on the device connections to maintain their continuity. If the continuity of the grounded conductor is interrupted, an unbalanced voltage condition could exist, creating higher voltages on parts of the multiwire branch circuit that may be damaging to equipment installed on those parts of the circuit. A wire nut connection with a "pigtail" on the grounded conductor is one potential solution.

51. What is the minimum length of free conductor that must extend outside the opening of a junction box having an opening of 8 inches in length by 8 inches in depth?

   a. 3 inches
   b. 6 inches
   c. 8 inches
   d. There is no requirement for the conductors to extend outside of the enclosure opening.

51. D: 300.14. In general, 6 inches of free conductor must be left at all outlets, junction boxes, and other such enclosures, measured from the point the conductor emerges from the raceway or cable in the box. Additionally, for enclosures having an opening less than 8 inches in any direction, 3 inches of free conductor must be able to extend outside the opening of the enclosure. For openings measuring more than 8 inches in all directions, there is no requirement for the conductors to be able to extend outside of the opening.

52. The paralleled conductors that comprise each ungrounded conductor, grounded conductor, neutral conductor, equipment grounding conductor, equipment bonding jumper, or supply-side bonding jumper shall comply with all of the following, EXCEPT:

   a. Be the same length.
   b. Consist of the same conductor size and material.
   c. Have the same insulation type and termination manner.
   d. Have the same insulation color.

52. D: 310.10(G)(2). When conductors are paralleled, they are required to be the same length, consist of the same conductor material, be the same size (circular mil area), have the same insulation type, and be terminated in the same manner. While it is certainly good practice, there is no requirement for the conductors to have the same insulation color, although requirements for conductor type identification would still apply.

53. Excluding exceptions, conductors in parallel shall not be permitted in sizes smaller than _____.

   a. 2 AWG
   b. 1 AWG
   c. 1/0 AWG
   d. 2/0 AWG

53. C: 310.10(G)(1). In general, conductors connected in parallel shall not be smaller than 1/0 AWG.

54. Determine the ampacity of three size 12 AWG copper type THHN current-carrying conductors installed in a raceway with an ambient temperature of 30 °C.

    a. 20 amperes
    b. 25 amperes
    c. 30 amperes
    d. 40 amperes

54. C: Table 310.16 is used for determining the ampacity of insulated conductors with not more than three current-carrying conductors in a raceway, cable, or directly buried. Table 310.16 is also based on an ambient temperature of 30 °C (86 °F). When these conditions change, electricians must adjust the ampacity of the conductors in accordance with 310.15(B) and 310.15(C). Temperature corrections and adjustment factors shall be permitted to be applied to the ampacity for the temperature rating of the conductor. Although electricians often base conductor sizing off of the temperature limitations of terminations and equipment, when adjusting for temperature or conductor bundling, they are permitted to do the adjustment based on the conductor temperature rating, i.e., the 90 °C column for THHN.

The conditions of use for this question do not require any additional adjustments to the conductor's ampacity because they are being used in accordance with the design of Table 310.16. Though electricians must always ensure that they do not exceed the temperature limitations at any point in the circuit, strictly speaking, the ampacity of 12 AWG copper type THHN conductors is 30 amperes.

55. Determine the ampacity of eight size 8 AWG copper type RHH current-carrying conductors installed in a raceway with an ambient temperature of 30 °C.

    a. 24.5 amperes
    b. 38.5 amperes
    c. 40 amperes
    d. 55 amperes

55. B: Table 310.16 is used for determining the ampacity of insulated conductors with not more than three current-carrying conductors in a raceway, cable, or directly buried. Table 310.16 is also based on an ambient temperature of 30 °C (86 °F). When these conditions change, electricians must adjust the ampacity of the conductors in accordance with 310.15(B) and 310.15(C). Temperature corrections and adjustment factors shall be permitted to be applied to the ampacity for the temperature rating of the conductor. Although conductor sizing is often based off of the temperature limitations of terminations and equipment, when adjusting for temperature or conductor bundling, electricians are permitted to do the adjustment based on the conductor temperature rating, i.e., the 90 °C column for THHN.

The conditions of use for the conductors in this question require an adjustment of the ampacity to compensate for the additional conductors beyond the design of Table 310.16. Table 310.15(C)(1) is used to determine the adjustment factor when more than three current-carrying conductors are installed together. Using Table 310.15(C)(1), an adjustment factor of 70% must be applied to the ampacity of the conductor. Based off of 310.16, the ampacity of 8 AWG copper RHH is 55 amperes.

$$55 \text{ A} \times 0.7 = 38.5 \text{ A}$$

56. Determine the ampacity of six size 10 AWG type THW-2 current carrying-conductors with six 10 AWG spare conductors installed in a raceway with an ambient temperature of 30 °C.

    a. 20 amperes
    b. 30 amperes
    c. 35 amperes
    d. 40 amperes

56. A: Table 310.16 is used for determining the ampacity of insulated conductors with no more than three current-carrying conductors in a raceway, cable, or directly buried. Table 310.16 is also based on an ambient temperature of 30 °C (86 °F). When these conditions change, electricians must adjust the ampacity of the conductors in accordance with 310.15(B) and 310.15(C). Temperature corrections and adjustment factors shall be permitted to be applied to the ampacity for the temperature rating of the conductor. Although conductor sizing is often based off of the temperature limitations of terminations and equipment, when adjusting for temperature or conductor bundling, electricians are permitted to do the adjustment based on the conductor temperature rating, i.e., the 90 °C column for THHN.

The conditions of use for the conductors in this question require electricians to adjust the ampacity to compensate for the additional conductors beyond the design of Table 310.16. Table 310.15(C)(1) is used to determine the adjustment factor when more than three current-carrying conductors are installed together. Though this application includes six current-carrying conductors, per the note on Table 310.15(C)(1), the six spare conductors must also be included in the adjustment. The adjustment factor for 12 current-carrying conductors in a raceway is 50%. Based on Table 310.16, the ampacity of 10 AWG copper THW-2 is 40 amperes.

$$40 \text{ A} \times 0.5 = 20 \text{ A}$$

57. Determine the ampacity of three size 3/0 AWG copper type THWN-2 current-carrying conductors in a raceway with an ambient temperature of 50 °C.

    a. 165.6 amperes
    b. 176.5 amperes
    c. 184.5 amperes
    d. 225 amperes

57. C: Table 310.16 is used for determining the ampacity of insulated conductors with no more than three current-carrying conductors in a raceway, cable, or directly buried. Table 310.16 is also based on an ambient temperature of 30 °C (86 °F). When these conditions change, electricians must adjust the ampacity of the conductors in accordance with 310.15(B) and 310.15(C). Temperature corrections and adjustment factors shall be permitted to be applied to the ampacity for the temperature rating of the conductor. Although base conductor sizing is often based off of the temperature limitations of terminations and equipment, when adjusting for temperature or conductor bundling, electricians are permitted to do the adjustment based on the conductor temperature rating, i.e., the 90 °C column for THHN.

The conditions of use for the conductors in this question require an adjusted ampacity to compensate for the increased ambient temperature beyond the design of Table 310.16. Table 310.15(B)(1) is used to determine the ambient temperature adjustment factor when the ambient temperature is not 30 °C. With an ambient temperature of 50 °C, the ampacity of the conductors must be adjusted by a factor of 0.82. Based on Table 310.16, the ampacity of 3/0 AWG copper THHN is 225 amperes.

$$225 \text{ A} \times 0.82 = 184.5 \text{ A}$$

58. Conductors shall not be _____ within a cable or cutout box unless a gutter having a width in accordance with Table 312.6(A) is provided. Conductors in parallel in accordance with 310.10(G) shall be judged on the basis of the number of conductors in parallel.
   a. Deflected
   b. Installed
   c. Terminated
   d. Looped

58. A: 312.6(A). The term "gutter" used in this section refers to the additional space used to provide the required bending/deflection space inside enclosures. What most electricians call a gutter is correctly referred to as a wireway. The intention of this section is to ensure that adequate "gutter space" exists in cabinets, cutout boxes, panels, etc., to allow for bending the conductors without damaging them.

59. Calculate the minimum wire-bending space required for the termination of 4/0 AWG ungrounded conductors, where the conductors enter the enclosure in the wall opposite the terminals where the conductors will be terminated.
   a. 4 inches
   b. 5 inches
   c. 6 inches
   d. 7 inches

59. D: 312.6(B) and Table 312.6(B). When determining the minimum wire-bending space required at each terminal, Table 312.6(B) is used. Using Table 312.6(B) reveals a minimum wire-bending space of 7 inches required for 4/0 AWG conductors with one wire per terminal. For certain installations having "lay-in terminals," such as meter sockets, Table 312.6(B) permits reductions to the minimum-wire bending space required at terminations per the notes at the bottom of Table 312.6(B).

60. Round boxes _____ where conduits or connectors requiring the use of locknuts or bushings are to be connected to the side of the box.

    a. Shall be used
    b. Shall not be used
    c. Shall be permitted
    d. None of the above

60. B: 314.2. Round boxes are not permitted where conduits or connectors require the use of a locknut or bushing and are connected on the side of the box. Bushings or locknuts used in this scenario are not able to make adequate contact with the entire surface of the box due to the curved walls. This does not allow for effective ground continuity.

61. Determine the box volume of a trade size 4-inch by 2⅛-inch box and a 4-inch square plaster ring marked with a 3.5-cubic inch volume.

    a. 25 cubic inches
    b. 30.3 cubic inches
    c. 33.8 cubic inches
    d. 45.5 cubic inches

61. C: 314.16(A) and Table 314.16(A). When calculating the total volume available within a box, the total volume of the assembled sections can be used, provided that they are marked with their volume or have their volume listed in Table 314.16(A). Plaster rings and other types of extensions, where their volumes are not marked, should not be considered when calculating the total volume of a box.

Volume of Components

4-inch x 2⅛-inch square box	= 30.3 cubic inches [Table 314.16(A)]
Plaster ring	= 3.5 cubic inches (marked on device)
Total volume	= 33.5 cubic inches

62. How many 10 AWG THHN conductors can fit in a trade size 4 and 11/16-inch by 1½-inch square box, where no other volume allowances are required?

    a. 10
    b. 11
    c. 12
    d. 16

62. B: 314.16(B)(1), Table 314.16(A), and Table 314.16(B)(1). The most basic box fill calculations involve either sizing a box to the number of conductors or determining the number of conductors of a specific size that can fit in a given box. Further box fill allowances for devices or items commonly installed in boxes such as receptacles, switches, and cable clamps are done by adding additional volume allowances in accordance with 314.6(B)(2), 314.6(3), 314.6(4), and 314.6(5). This question simply asks how many 10

AWG conductors can fit in a 4 and 11/16-inch by 1½-inch square box. The question states that no other volume allowances will be needed.

4 and 11/16-inch by 1½-inch square box = 29.5 cubic inches [Table 314.16(A)]
Volume allowance for 10 AWG conductor = 2.5 cubic inches [Table 314.16(B)(1)]

$$\frac{29.5 \text{ cubic inches}}{2.5 \text{ cubic inches}} = 11.8 \text{ or } 11 \text{ cubic inches}$$

63. How many cables containing three 2 AWG insulated THHN copper conductors and one 6 AWG bare copper conductors can be installed in a 12-inch ventilated trough cable tray if the outside diameter of each cable 1¼ inches?

    a. 9
    b. 10
    c. 11
    d. 12

63. C: 392.22(A)(1)(b) and Table 392.22(A)(1). When sizing cable trays for multiconductor cables smaller than 4/0 AWG, Table 392.22(A)(1), Column 1 is used to find the allowable cable fill for each cable tray size. Note that the sum of the cross-sectional area of the cables is used as opposed to the outside diameter.

$$A = \pi r^2$$
$$A = 3.14(0.625^2)$$
$$A = 1.227 \text{ square inches}$$

Allowable fill of a 12-inch ventilated cable tray = 14 square inches

$$\frac{14 \text{ square inches}}{1.227 \text{ square inches}} = 11.41 \text{ square inches}$$

or an 11-inch cable tray.

64. What size solid bottom cable tray would be required for the installation of eight 350 kcmil multiconductor cables having an outside diameter of 2½ inches? No additional spacing is required between cables.

    a. 20-inch
    b. 24-inch
    c. 30-inch
    d. 36-inch

64. B: 392.22(A)(3)(a). Where installed in a ventilated cable tray and all cables are 4/0 AWG or larger, the sum of the outside diameter of all cables installed must not exceed the width of the cable tray and must be installed in a single layer. Additionally, when installed in a solid bottom cable tray, as in this question, the sum of the outside diameter of the cables shall not exceed 90% of the width of the cable tray and must be installed in a single layer.

Eight cables that are 2½ inches in diameter = 20 inches.

$$\frac{20 \text{ inches}}{0.9} = 22.22 \text{ inches}$$

or 24-inch cable tray.

ELECTRICAL EQUIPMENT AND DEVICES

65. The conductor intended to be used as the grounded conductor of a flexible cord or fixture wire shall be identified by any of the following methods, EXCEPT:

    a. One conductor having the individual strands tinned and the other conductor(s) having the individual strands untinned for cords having insulation on the individual conductor's integral with jacket.
    b. One or more ridges, grooves, or white stripes located on the exterior of the cord so as to identify one conductor for cords having insulation on the individual conductor's integral with the jacket.
    c. An embossed or engraved marking within the individual conductor's insulation bearing the words "grounded" or "neutral" at least every 12 inches along the length of the entire exposed section of wiring.
    d. A braid finished to show a white or gray color and the braid on the other conductor(s) finished to show a readily distinguishable solid in color or colors.

65. C: 400.22(A)-(F). The conductor intended to be used as the grounded circuit conductor for both flexible cords and fixture wires can be identified by any means listed in 400.22(A)-(F). The nearly universal method for identifying the grounded conductor on fixture wires is the use of ridges or grooves along the length of the insulation on the grounded conductor. Care must be taken to correctly identify the grounded conductor to prevent the application of voltage on the screw shell of the lampholder.

66. In general, switches or circuit breakers shall not disconnect the _____ conductor of a circuit.

    a. Grounded
    b. Ungrounded
    c. White
    d. Both a and b

66. D: 404.2(B), 200.7(C). Except in the rare exception outlined in 404.2(B), switches and circuit breakers shall not disconnect the grounded conductor of a circuit. Also of note, 200.7(C) states that conductors having white, gray, or striped insulation shall not be used as the return or "switch leg," but are allowed to be used as the supply to a switch if properly reidentified to indicate its use as an ungrounded conductor by marking tape or other approved means of a color other than white, gray, or green.

67. A single snap switch in a single-family dwelling controls four recessed lighting fixtures whose lampholders are marked with a maximum wattage of 75 watts. The conductors between the overcurrent protection, the switch, and the luminaires are 12 AWG copper. The overcurrent protection device is rated 20 amperes at 120 volts. What is the minimum required rating of the switch?

    a. 15 amperes
    b. 20 amperes
    c. 25 amperes
    d. 30 amperes

67. A: 404.14(A). Though the overcurrent protection device protecting the switch and the conductors supplying the switch are rated 20 amperes, 404.14(A) states that for resistive, LED, CFL, and other types of lamps that are commonly installed in such fixtures, the load of the switch must not be exceeded.

75 W max × 4 lamps = 300 W total

$$\frac{300\text{ W}}{120\text{ V}} = 2.5\text{ A}$$

2.5 amperes is well below the 15-ampere rating of the switch. Therefore, a 15-ampere switch is all that is required for this installation.

68. Receptacles incorporating an isolated equipment grounding conductor connection shall be identified by _____.

    a. A green circle on the face of the receptacle
    b. An orange triangle on the face of the receptacle
    c. A green triangle on the face of the receptacle
    d. The letters "IG" or the words "isolated ground" on the face of the receptacle

68. B: 406.3(E). Isolated ground receptacles shall be identified by the use of an orange triangle on the face of the receptacle.

69. When replacing non-grounding type receptacles where an existing equipment grounding conductor does not exist, which of the following are required?

    a. Replace with another non-grounding type receptacle
    b. Replace with a ground-fault circuit-interrupter-type receptacle. The receptacles or their covers must be marked "no equipment ground." The equipment grounding conductor shall not be connected from the ground-fault circuit-interrupter-type receptacle to any outlet supplied from the ground-fault circuit-interrupter receptacle.
    c. Replace with a grounding-type receptacle where supplied through a ground-fault circuit-interrupter and marked "GFCI protected" and "no equipment ground" visible after installation. An equipment grounding conductor shall not be connected between the grounding-type receptacles.
    d. Any of the above

69. D: 406.4(D)(2). Prior to the 1960s, there were few requirements for providing a grounding means for residential and commercial branch circuits supplying cord-and-plug connected loads. Branch circuit wiring that does not contain an equipment grounding conductor and two-wire, non-grounding receptacles can still be found in many older homes. When replacing these receptacles, electricians must meet the requirements of 406.4(D). Installation of a grounding-type receptacle on a circuit where a proper equipment ground is not present introduces a significant safety hazard because it essentially guarantees that any ground fault that may develop would not be provided with the low resistance path to ground that allows the overcurrent device to trip. This leaves the fault searching for alternative paths or, even worse, energized metal parts that normally do not carry a current, making a serious electrocution hazard out of typically harmless objects.

70. After installation, receptacles shall be installed so that _____.
   a. The face of the receptacle is flush with or projects out from faceplates of insulating material and must project at least 0.015 inches from metal faceplates
   b. The receptacle is oriented with its grounding slot up
   c. The receptacle is oriented with its grounding slot down
   d. The terminals are at least a ½-inch from any metallic parts of its enclosure

70. A: 406.5(D). Much debate exists about the correct orientation of vertically installed receptacles. Contrary to the declarations of many confident electricians, the NEC does not make any distinction as to whether receptacles should be installed "ground-up" or "ground-down." 406.5(D) does however state that the face of receptacles should be at least flush with the faceplate when the faceplate material is insulating material, and must project at least 0.015 inches past the faceplate when the faceplate is metal.

71. Receptacles of 15 or 20 amperes, 125 and 250 volts installed in a wet location shall_____.
   a. Have an enclosure that is weatherproof when the receptacle is covered (attachment plug cap not inserted and receptacle covers closed)
   b. Have an enclosure that is weatherproof whether or not the attachment plug cap is inserted and shall be listed and identified as "extra-duty" where installed with an outlet box hood
   c. Be of the weather-resistant type and have an enclosure that is weatherproof when the receptacle is covered (attachment plug cap not inserted and receptacle covers closed)
   d. Be of the weather-resistant type and have an enclosure that is weatherproof whether or not the attachment plug cap is inserted and shall be listed and identified as "extra-duty" where installed with an outlet box hood

71. D: 406.9(B) Receptacles installed in wet locations must be of the weather-resistant type, and must have an enclosure that is weatherproof even when there is a cord cap plugged in. These weatherproof covers often make use of a large hood to allow the cord to be plugged into the receptacle while still allowing the cord to be safely bent and exit out of the bottom. Often referred to as "in-use" covers, the hoods must be listed and identified as "extra-duty."

72. All 15- and 20-ampere, 125- and 250-volt non-locking receptacles installed in childcare facilities must be _____.

    a. Self-grounding type receptacles
    b. Isolated ground receptacles
    c. Tamper-resistant receptacles
    d. Hospital grade receptacles

72. C: 406.12 lists all areas where tamper-resistant receptacles are required. Tamper-resistant receptacles are constructed with shutters to prevent children from inserting objects into the receptacle slots. The shutters will only open when equal pressure is applied to on both sides.

73. Switchboards, switchgear, and panelboards shall have a short-circuit current rating not less than _____.

    a. The available fault current
    b. 50 kA
    c. 100 kA
    d. 200 kA

73. A: 408.6. The short-circuit current rating (SCCR) of any switchboard, switchgear, or panelboard must be at least as great as the available fault current. Equipment having an SCCR greater than or equal to the available fault current ensures that the equipment can handle any possible faulting condition without failing. In other than one- and two-family dwelling units, the available fault current and the date of calculation must be field marked on the enclosure.

74. Plug-in-type overcurrent protection devices or plug-in type main lug assemblies that are backfed and used to terminate field-installed ungrounded supply conductors shall _____.

    a. Be secured in place by an additional fastener that requires a pull-to-release the device from the mounting means on the panelboard
    b. Be secured in place by an additional fastener that requires *other than* a pull-to-release the device from the mounting means on the panelboard
    c. Not have a rating less than 100 amperes
    d. Not be allowed

74. B: 408.36(D). Plug-in type breakers used as the disconnecting means or overcurrent protection for a panelboard must be fastened in a manner other than the typical pull-to-release action used with plug-in-type breakers.

75. A luminaire installed within the actual outside dimensions of a bathtub or shower to a height of 8 feet vertically from the top of the bathtub rim or shower threshold and not subject to shower spray shall be _____.
   a. Marked suitable for damp locations
   b. Marked suitable for wet locations
   c. Either a or b
   d. A wet-niche luminaire

75. C: 410.10(D)(2). Luminaires installed within the footprint of a bathtub or shower and up to a height of 8 feet above the bathtub rim or shower stall must be marked either suitable for damp locations or suitable for wet locations. A luminaire installed in such a location and subject to shower spray must be marked suitable for wet locations. Wet-niche luminaires are a type of light fixture used for underwater applications such as pool lights, but are not a requirement for bathtubs or showers.

76. Luminaires installed in clothes closets shall be permitted to be any of the following types, EXCEPT:
   a. Surface-mounted or recessed incandescent or LED luminaires with completely enclosed light sources.
   b. Incandescent luminaires with open or partially enclosed lamps.
   c. Surface-mounted or recessed fluorescent luminaires.
   d. Surface-mounted fluorescent or LED luminaires identified as suitable for installation within the clothes closet storage space.

76. B: 410.16(A). Sometimes referred to as "the most overlooked fire hazard in a home," an open incandescent bulb can reach temperatures sufficient to ignite the flammable materials commonly found in many clothes closets. The NEC requires the use of totally enclosed incandescent or LED luminaires, surface-mounted or recessed fluorescent luminaires, or luminaires identified as suitable for installation within the clothes closet storage space.

77. A recessed luminaire is installed within a ½-inch of combustible material and will be covered with thermal insulation. This luminaire must be a _____ luminaire.
   a. LED
   b. Fluorescent
   c. Non-type IC
   d. Type IC

77. D: 410.116(A). A recessed luminaire that cannot be spaced more than a ½-inch from combustible materials and cannot maintain 3 inches of clearance from thermal insulation around its enclosure, wiring compartment, ballast, transformer, LED driver, or power supply must be of the type IC (insulation contact). Though LED and fluorescent luminaires produce significantly less radiated heat from their light source, the ballasts and power supplies of such luminaires can still produce enough heat to pose a fire hazard, and thus must also be Type IC.

MOTORS AND GENERATORS

78. Determine the ampacity as calculated by the NEC for the purpose of determining the ampacity of conductors, ampere rating of switches, or branch-circuit short-circuit and ground-fault protection for a 20 horsepower, three-phase, 460-volt induction-type squirrel cage motor having a nameplate full-load amperage of 23 amperes.

  a. 23 amperes
  b. 27 amperes
  c. 34 amperes
  d. 33.75 amperes

78. B: 430.6(A)(1). For the purpose of determining the ampacity of motors other than low speed, high torque, or multispeed motors, the values given in Table 430.247 (DC motors), Table 430.248 (single-phase AC motors), Table 430.249 (two-phase AC motors), and Table 430.250 (three-phase AC motors) are to be used when calculating the ampacity of conductors, current rating of switches, and current ratings of branch-circuit short-circuit and ground-fault protection, as opposed to using the ampere rating on the nameplate of the motor for the calculations. When a motor is marked in amperes, but not horsepower, the horsepower rating shall be assumed to correspond with the value given in the tables above.

79. Determine the minimum size copper THHN motor circuit conductors required to supply a single 50-horsepower, three-phase, 230-volt squirrel cage motor.

  a. 1 AWG
  b. 1/0 AWG
  c. 2/0 AWG
  d. 3/0 AWG

79. C: 430.6(A)(1), 430.22, Table 430.250, and Table 310.16. Conductors that supply a single motor used in a continuous duty application must have an ampacity not less than 125% of the full-load current of the motor. 430.6(A)(1) requires that Table 430.250 be used to determine this ampacity based on the motor type, motor voltage, and horsepower. Table 430.250 gives an ampacity of 130 amperes for a 50 horsepower, three-phase squirrel-cage motor operating at 230 volts.

$$130 \text{ A} \times 1.25 = 162.5 \text{ A}$$

Table 310.16 is used to determine the ampacity of conductors when no more than three current-carrying conductors are in a raceway. THHN conductors have a max temperature rating of 90 °C. However, these temperatures would exceed the likely temperature limitations of the terminations as stated in 110.14(C). The vast majority of terminations, both on circuit breakers and motor leads, are rated 75 °C max. Therefore, based on the 75 °C column of Table 310.16, 2/0 AWG copper is required.

80. Determine the minimum size copper THWN conductors required to supply one 20-horsepower, three-phase, 460-volt induction motor; and two 10-horsepower, three-phase, 460-volt induction motors.

   a. 6 AWG
   b. 8 AWG
   c. 10 AWG
   d. 12 AWG

80. A: 430.24, Table 430.250, Table 310.16. Conductors that supply multiple motors, or a motor and additional loads, must have an ampacity not less than 125% of the full-load current of the largest motor, plus the sum of the full-load currents of all other motors in a group. 430.6(A)(1) requires that Table 430.250 be used to determine this ampacity based on the motor type, motor voltage, and horsepower. Additionally, if the conductors supply other non-motor loads, 100% of the non-continuous non-motor loads and 125% of the continuous non-motor loads must be added to the total motor load calculated previously.

Largest Motor:

20 hp, 460 V Motor $=$ 27 A (per Table 430.250)

27 A $\times$ 1.25 = 33.75 A

Other Motors:

10 hp, 460 V Motor $=$ 14 A (per Table 430.250)

14 A $\times$ 2 Motors $=$ 28 A

Total Calculated Load:

33.75 A + 28 A = 61.75 A

Table 310.16 is used to determine the ampacity of conductors when no more than three current-carrying conductors are in a raceway. THWN conductors have a max temperature rating of 75 °C. Therefore, based on the 75 °C column of Table 310.16, 6 AWG copper is required.

81. Separate motor overload protection shall be based on _____.

   a. Table 430.247, Table 430.248, Table 430.249, or Table 430.250
   b. The motor nameplate current rating
   c. The larger of a or b
   d. The smaller of a or b

81. B: 430.6(A)(2). When determining the setting of separate motor overload protection, the motor nameplate current rating is used. This is in contrast to the motor circuit conductors, short-circuit and ground-fault protection, and switch ratings, which are determined by the Tables in Article 430 based on the type of motor and its operating voltage and horsepower. It is also worth noting the difference between motor and branch-circuit overload protection and motor branch-circuit short-circuit and ground-fault protection. Induction motors inherently draw a large amount of inrush current during startup, often six to eight times that of the normal running current. This requires that the overcurrent protection device be sized large enough to withstand this inrush current without tripping. However, this leaves the both the motor and the motor circuit conductors unprotected from small overloads above their designed limits but below the trip rating of the overcurrent protection device. This is the purpose of a separate overload device. The device must be responsive to motor current and be selected to trip at a specific current based on the nameplate of the motor. In summary, *motor branch-circuit short-circuit and ground-fault protection* protects against line-to-line and line-to-ground faults within the motor circuit and motor. *Motor overload protection* protects against overloads above the designed limitations of the motor circuit conductors and motor, but below the ratings of the branch-circuit short-circuit and ground-fault protection.

82. Determine the trip setting of an overload device installed to protect a 5 horsepower, single-phase motor operating at 230 volts. The nameplate ratings are as follows: Full-load current is 32 amperes, service factor is 1.15, the temperature rise is 40 °C.

 a. 32 amperes
 b. 35 amperes
 c. 37 amperes
 d. 40 amperes

82. D: 430.6(A)(2), 430.32(A)(1). *Motor overload protection* protects against overloads above the designed limitations of the motor circuit conductors and motor, but below the ratings of the branch-circuit short-circuit and ground-fault protection. When determining the setting of separate overload devices, several nameplate ratings must be considered. 430.32(A)(1) requires that the setting be based on 125% of the full-load current as marked on the nameplate of the motor for motors having a service factor of 1.15 or greater and a marked temperature rise of 40 °C or less. For all other motors, the setting shall be based on 115% of the full-load current.

Nameplate Full-load Current = 32 amperes.

32 A × 1.25 = 40 A

83. A 10 horsepower, three-phase induction motor operating at 460 volts is equipped with a separate overload protection device. The nameplate on the motor states that the motor's full-load current is 15 amperes and its service factor is 1.0. The overload device is currently set to trip at 17.25 amperes, but the overload device frequently trips on motor startup. What changes to the current setting, if any, should be made to the trip setting on the overload device to prevent nuisance tripping on startup?

    a. No change should be made
    b. 18.75 amperes
    c. 19.5 amperes
    d. 21 amperes

83. C: 430.6(A)(2), 430.32(C). When an overload device or its setting is selected in accordance with 430.32(A)(1) and that setting is not sufficient to start the motor or to carry the load, 430.32(C) allows the setting to be increased to a percentage of the motor's full-load current, based on several nameplate ratings.

Motors with marked service factor 1.15 or greater: 140%

Motors with a marked temperature rise of 40 °C or less: 140%

All other motors: 130%

Due to the service factor of the motor being less than 1.15 (1.0), the maximum trip setting allowed by this section is 130% of the motor's full-load current (15 amperes).

$$15 \text{ A} \times 1.3 = 19.5 \text{ A}$$

ELECTRICAL CONTROL DEVICES AND DISCONNECTING MEANS

84. The circuit of a control apparatus or system that carries the electrical signals directing the performance of the controller but does not carry the main power current best describes _____.

    a. A controller
    b. A disconnecting means
    c. A control circuit
    d. A power-limited circuit

84. C: Article 100. Control circuits, though sometimes derived directly from the source of power supplying the device being controlled, are defined as directing the performance of the controller without carrying the main power itself.

85. A device, group of devices, or other means by which the conductors of a circuit can be disconnected from their source of supply best describes _____.

    a. Disconnecting means
    b. A transfer switch
    c. Switchgear
    d. Bypass isolation switch

85. A: Article 100. Disconnecting means are devices that disconnect the conductors from their source of supply. Notably, by definition disconnecting means do not provide any overcurrent protection, though there are scenarios when an overcurrent protection device can serve as a disconnecting means.

86. For fixed outdoor deicing and snow melting equipment readily accessible to the user of the equipment, the _____ shall be permitted to serve as the disconnecting means when it is of the indicating type and capable of being locked in the open (off) position.

    a. Branch-circuit switch
    b. Circuit breaker
    c. Factory-installed attachment plug rated 20 amperes or less and 150 volts to ground or less
    d. Any of the above

86. D: 426.50(A), 426.50(B). As is the case in many installations, when the branch-circuit overcurrent protection device meets certain requirements, i.e., an indicating type and capable of being locked in the off positions, it is allowed to serve as the disconnecting means, often serving the purpose of a maintenance disconnect. Similarly, the plug of cord-and-plug connected equipment is occasionally permitted to serve as the disconnecting means.

87. For indoor locations other than dwelling units and associated structures where existing luminaires are installed without a disconnecting means, at the time a ballast is replaced, a(n) _____ shall be installed.

    a. In-line overcurrent protection device
    b. Disconnecting means
    c. Ground-fault circuit-interrupter (GFCI)
    d. Surge protection device

87. B: 410.71(G)(1). Fluorescent and LED luminaires are required to have a disconnecting means, either internal or external to the fixture, to allow for the safe removal of power without having to turn all the lights off, leaving the working area in total darkness, or for the replacement of the ballast or driver. If a luminaire is found to have been installed without a disconnecting means, one should be installed during ballast replacement. These normally take the form of factory-installed quick-disconnects internal to the luminaire near the ballast or driver.

SPECIAL OCCUPANCIES, EQUIPMENT, AND CONDITIONS

88. Locations in which flammable gases, flammable liquid-produced vapors, or combustible liquid-produced vapors are or may be present in the air in quantities sufficient to produce explosive or ignitable mixtures best describe what type of location?

    a. Class I
    b. Class II
    c. Class III
    d. None of the above

88. A: 500.5(B). Class I locations are those in which flammable gases, flammable liquid-produced vapors, or combustible-liquid produced vapors are or may be present in the air in sufficient quantities to produce an explosive or ignitable mixture. The NEC is not responsible for making a determination as to whether a specific installation meets the definition of a Class I location.

89. A _____ location is a location in which ignitable concentrations of flammable gases, flammable liquid-produced vapors, or combustible liquid-produced vapors can exist under normal operation, may exist frequently because of repair or maintenance, or may exist due to the breakdown or faulty operation of equipment and might also cause simultaneous failure of electrical equipment in such a way as to directly cause the electrical equipment to become a source of ignition.

    a. Class I, Division 1
    b. Class I, Division 2
    c. Class II, Division 1
    d. Class II, Division 2

89. A: 500.5(B)(1). Class I locations are separated into two divisions, Division 1 and Division 2. Division 1 locations are areas where ignitable concentrations of gases, liquids, or vapors exist under normal operation or are likely to exist during routine maintenance. Additionally, if the faulty operation or breakdown of equipment might also cause the simultaneous failure of electrical equipment, then that could then become the source of ignition. Division 2 locations are areas where a failure of containment vessels; mechanical ventilation; or the proximity to a Class I, Division 1 location might occasionally lead to ignitable concentrations of combustible liquids, vapors, or gases.

90. _____ locations are those that are hazardous because of the presence of combustible dust.

    a. Class I
    b. Class II
    c. Class III
    d. None of the above

90. B: 500.5(C). Class II locations are those in which the primary hazard is due to the presence of combustible dust. The NEC is not responsible for making a determination as to whether a specific installation meets the definition of a Class II location.

91. A _____ location is a location in which combustible dust is in the air under normal operating conditions and where mechanical failure or abnormal operations of machinery or equipment might cause such explosive or ignitable mixtures to be produced while the simultaneous failure of electrical equipment could become a source of ignition, or in which Group E combustible dust may be present in quantities sufficient to be hazardous in normal or abnormal operating conditions.

    a. Class I, Division 1
    b. Class I, Division 2
    c. Class II, Division 1
    d. Class II, Division 2

91. C: 500.5(C)(1). Class II locations are separated into two divisions: Division 1 and Division 2. Division 1 locations are areas where combustible dust in the air under normal operating conditions exists in quantities sufficient to produce explosive or ignitable material if the faulty operation or breakdown of equipment might also cause the simultaneous failure of electrical equipment that could then become the source of ignition. Also, if Group E combustible dust such as magnesium or aluminum are present, they can be particularly dangerous due to the tendency of current flow or an electrical arc causing ignition. Division 2 locations are areas where combustible dust due to abnormal operations may be present in the air, or accumulations are present but are normally insufficient to interfere with the normal operation of electrical equipment or other apparatus, but could interfere with the safe dissipation of heat from electrical equipment or could be ignitable by abnormal operation or failure of electrical equipment as a result of infrequent malfunctioning of handling or processing equipment; or in which combustible dust accumulations on, in, or in the vicinity of the electrical equipment.

92. Locations that are hazardous because of the presence of easily ignitable fibers, or where materials producing combustible fibers or flyings are handled, manufactured, or used, but in which such fibers/flyings are not likely to be suspended in the air in quantities sufficient to produce ignitable mixtures are best described as what type of location?

    a. Class I
    b. Class II
    c. Class III
    d. Class IV

92. C: 500.5(D). Class III locations are those in which the primary hazard is due to the presence of easily ignitable fibers or flyings. Such locations include textile mills and similar manufacturing facilities and frequently involve materials such as rayon and cotton. The NEC is not responsible for making a determination as to whether a specific installation meets the definition of a Class III location.

93. An explosion-proof enclosure containing a 20-ampere, single-pole switch is installed in a Class I, Division 1 location. A seal must be installed within _____ of the enclosure.

    a. 12 inches
    b. 18 inches
    c. 24 inches
    d. A seal is not required for this installation

93. B: 501.15(A)(1). In Class I, Division 1 locations, explosion-proof enclosures containing devices such as switches, circuit breakers, fuses, relays, or resistors that may produce arcs, sparks, or temperatures exceeding 80% of the autoignition temperature of the gas or vapor must have seals installed within 18 inches of the enclosure. When the switch, circuit breaker, fuse, relay, or resistor is enclosed within a chamber hermetically sealed against the entrance of gases or vapors and the conduit is less than trade size 2 inches, no seal is required.

94. A trade size 1-inch rigid metal conduit passes completely through a Class I, Division 2 location and terminates at a point 15 inches inside an unclassified location. The portion of the conduit that traverses the Class I, Division 2 location contains no unions, couplings, boxes, or fittings. A seal is required within _____ of the enclosure.

    a. 12 inches
    b. 18 inches
    c. 24 inches
    d. A seal is not required for this installation

94. D: 501.15(B)(2) Exception No. 1. Metal conduits passing completely through Class I, Division 2 locations and not containing unions, couplings, boxes, or fittings that also do not have fittings installed within 12 inches of either side of the boundary shall not be required to be sealed.

95. Which of the following is NOT a permitted wiring method for Class II, Division 1 locations?

    a. Threaded rigid metal conduit
    b. Threaded steel intermediate metal conduit
    c. Electrical metallic tubing
    d. Type MI cable with listed fittings supported to avoid tensile stress at terminations

95. C: 502.10(A)(1). In all hazardous locations, special consideration must be given to the wiring methods regarding the nature of the hazard and its ability to intrude and communicate through the electrical installation via the wiring method. Conduits often terminate at enclosures or devices that contain sparking, arcing, or heat-generating devices. Wiring methods must be chosen based on their ability to resist the intrusion of the specific hazardous material for the particular hazardous location. The wiring method requirements for each class will be found in the 50X.10 section of the relevant article.

96. A space in which failure of a system or equipment is likely to cause minor injury to patients, staff, or visitors is best described as a _____.

- a. Category 1 (critical care) space
- b. Category 2 (general care) space
- c. Category 3 (basic care) space
- d. Category 4 (support) space

96. B: Article 100. In previous versions of the NEC, patient care spaces were divided into four areas: critical care rooms, general care rooms, basic care rooms, and support rooms. The 2023 NEC uses the wording Category 1, Category 2, Category 3, and Category 4 spaces, with Category 1 spaces being the most likely to cause injury or death and Category 4 spaces being the least likely to cause injury or death.

97. Patient bed locations in Category 2 spaces shall be supplied by at least two branch circuits and shall be provided with a minimum of _____ receptacles.

- a. 4
- b. 6
- c. 8
- d. 10

97. C: 517.18(B)(1). Receptacles in Category 2 spaces (general care), such as inpatient bedrooms, dialysis rooms, etc., shall be provided with a minimum of eight receptacles. Additionally, the Category 2 space must be supplied by at least two branch circuits, one from the normal branch and one from the critical branch. For the purpose of this code section, a duplex receptacle is considered two receptacles.

98. All receptacles installed in patient bed locations in Category 1 and Category 2 spaces shall be listed as _____ receptacles.

- a. Hospital-grade
- b. Isolated ground
- c. Self-grounding
- d. Tamper-resistant

98. A: 517.18(B)(2), 517.19(B)(2). Other than the exceptions to these code sections, all receptacles installed in patient bed locations of Category 1 and Category 2 spaces must be of the "hospital-grade" type. Hospital-grade receptacles are usually identified by a green dot on the face of the receptacle. Hospital-grade receptacles are required to meet additional performance requirements. These include grounding reliability, assembly integrity, strength, and durability tests.

RENEWABLE ENERGY TECHNOLOGIES

99. Determine the maximum circuit current of five parallel connected photovoltaic modules, each having a short circuit current of 9.75 amperes.

    a. 48.75 amperes
    b. 50 amperes
    c. 60.93 amperes
    d. 70 amperes

99. C: 690.8(A)(1). Photovoltaic modules, also known as solar panels, are electrical power-producing devices. Each module will produce a known maximum current if the output of the module is shorted during periods of high production, i.e., solar noon. It is permissible to determine the maximum circuit current that the circuit conductors of parallel connected modules would be subject to by adding the short circuit currents of each module in parallel and multiplying the sum by 125%.

Short circuit current of each module = 9.75 amperes

9.75 A × 5 modules = 48.75 A

48.75 A × 1.25 = 60.93 A

100. Where disconnecting means of a photovoltaic system above 30 volts are readily accessible to unqualified persons, any enclosure door or hinged cover that exposes live parts when open shall _____.

    a. Be locked
    b. Require a tool to open
    c. Both a and b
    d. Either a or b

100. D: 690.13(A)(2). Photovoltaic systems must have a disconnecting means capable of disconnecting the PV system from all wiring and systems, including power systems, energy storage systems, and utilization equipment and its associated premise wiring. This disconnecting means must be readily accessible. If it is readily accessible to unqualified persons, it must either be locked or require a tool to open.

Practice Test

1. A device intended for the detection of ground-fault currents, used in circuits with voltage to ground greater than 150 volts, that de-energizes a circuit when the ground-fault current exceeds the values established for Class C, D, or E devices best describes a:

    a. Circuit breaker
    b. AFCI
    c. SPGFCI
    d. GFCI

2. Distribution cutouts rated over 1,000 volts nominal shall not be used _____.

    a. Outdoors, aboveground, or in non-metallic enclosures
    b. Indoors, underground, or in metal enclosures
    c. At less than their maximum rated voltage
    d. Where the fused cutouts are interlocked with a switch to prevent opening the cutouts under load

3. A piece of equipment operating at 23 kV nominal is being placed outdoors in a field-fabricated installation. Determine the minimum air separation between any of its bare live conductors and the ground.

    a. 10.5 inches
    b. 15 inches
    c. 7.5 inches
    d. 10 inches

4. Determine the minimum clearance for overhead service conductors crossing over a residential pool. This measurement is the distance from the water surface to the 120 volts to ground overhead service conductors.

    a. 10 feet
    b. 14.5 feet
    c. 18 feet
    d. 22.5 feet

5. Determine the minimum number of receptacles required to be installed to serve countertop appliances in a kitchen island that has a rectangular countertop measuring 3 feet wide by 8 feet long.

    a. 1
    b. 2
    c. 3
    d. There is no requirement to install receptacles for island countertops

6. Where installed in a metal raceway or enclosure, all conductors of all feeders using a common neutral conductor shall be _____.

    a. Enclosed within a listed electromagnetic interference (EMI) cable shielding jacket

    b. Wrapped with a wire mesh tape where conductors are routed near metal parts of any enclosure

    c. Both a and b

    d. Enclosed within the same raceway or other enclosure as required in 300.20

7. Determine the maximum standard ampere rating of an overcurrent protection device for a 1,000 VA control transformer with a single-phase 480-volt primary and single-phase 120-volt secondary. The control transformer will have primary protection only.

    a. 1

    b. 3

    c. 6

    d. 10

8. Two dual element (time-delay) type RK5 fuses are used as the motor branch-circuit short-circuit and ground-fault protection for a 7½ horsepower, single-phase 208-volt motor. What is the maximum allowable current rating of these fuses?

    a. 70 amperes

    b. 80 amperes

    c. 110 amperes

    d. 125 amperes

9. An emergency power source used for emergency illumination shall be of suitable rating and capacity to supply and maintain the total load for the duration determined by the system design, but not less than _____.

    a. 1 hour

    b. 1½ hours

    c. 2 hours

    d. 2½ hours

10. Determine the length of a straight pull box where the conduit entering the box is a 3-inch EMT containing four 4/0 THHN conductors.

    a. 16 inches

    b. 24 inches

    c. 30 inches

    d. 32 inches

11. For health care facilities, which of the following branches are not permitted to be arranged for delayed automatic connection to the alternate power source?

    a. Central suction system serving medical and surgical functions

    b. Compressed air systems serving medical and surgical functions

    c. Exit signs

    d. Smoke control and stair pressurization systems

12. All wind turbines shall be required to have a readily accessible manual shutdown button or switch, unless they have a swept area less than _____.

   a. 500 square feet
   b. 538 square feet
   c. 560 square feet
   d. 583 square feet

13. Which of the following is NOT true for elevator hoistway pit lighting and receptacles?

   a. A separate branch circuit shall supply the hoistway pit lighting and receptacles.
   b. The lighting switch shall be located so as to be readily accessible from the pit access door.
   c. Hoistway pit lighting shall be connected to the load side of a ground-fault circuit interrupter.
   d. At least one 125-volt, single-phase, 15- or 20-ampere duplex receptacle shall be provided in the hoistway pit.

14. Determine the number of wall space receptacles required to be installed in the room shown in the figure below.

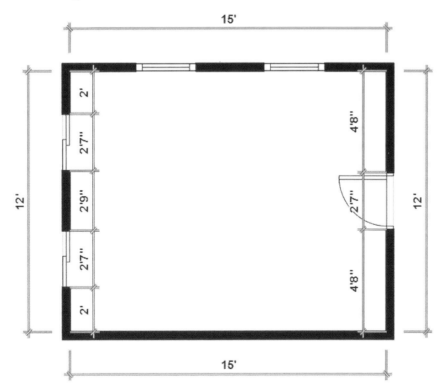

   a. 3
   b. 5
   c. 6
   d. 7

15. Where a portable generator is the source of an optional standby system and is used as a non-separately derived system, the equipment grounding conductor shall be bonded to the system _____.

    a. Grounding electrode
    b. Grounded conductor
    c. Main bonding jumper
    d. Any of the above

16. What is the maximum spacing between supports for trade size 2-inch rigid polyvinyl chloride conduit (PVC)?

    a. 3 feet
    b. 5 feet
    c. 6 feet
    d. 7 feet

17. Circuit breakers and switches containing fuses shall be located so that they may be operated from a readily accessible place. They shall be installed such that the center of the grip of the operating handle of the switch or circuit breaker, when in its highest position, is not more than _____ above the floor or working platform.

    a. 6 feet
    b. 6 feet 6 inches
    c. 6 feet 7 inches
    d. 7 feet

18. Capacitor cases shall be connected to the _____.

    a. Equipment grounding conductor
    b. Grounding electrode conductor
    c. Grounded conductor
    d. Ungrounded conductor

19. Each disconnecting means shall be legibly marked to indicate its purpose unless located and arranged so the purpose is evident. In _____, the marking shall include the identification and location of the circuit source that supplies the disconnecting means unless located and arranged so the identification and location of the circuit source is evident.

    a. One-family dwellings
    b. Two-family dwellings
    c. Other than one- or two- family dwellings
    d. Industrial settings

20. Determine the minimum vertical clearance above finished grade required for overhead service conductors over a commercial driveway subject to truck traffic.

    a. 10 feet
    b. 12 feet
    c. 15 feet
    d. 18 feet

21. A wall space is any space _____ or more in width (including space measured around corners) and unbroken along the floor line by doorways and similar openings, fireplaces, stationary appliances, and fixed cabinets that do not have countertops or similar work surfaces.

    a. 2 feet
    b. 3 feet
    c. 5 feet
    d. 6 feet

22. Floor receptacles located within _____ of the wall may be counted as part of the required number of wall space receptacle outlets.

    a. 6 inches
    b. 10 inches
    c. 12 inches
    d. 18 inches

23. For Type IMC, there shall not be more than _____ between pull points such as conduit bodies and boxes.

    a. 360 degrees of bends
    b. 270 degrees of bends
    c. 100 feet
    d. 200 feet

24. For rides, tents, and concessions of carnivals, circuses, fairs, and similar events, a means to disconnect each portable structure from all ungrounded conductors shall be provided within both sight and _____ of the operator's station.

    a. 6 feet
    b. 10 feet
    c. 15 feet
    d. 25 feet

25. The branch circuit conductors for fixed electric space heating equipment and any associated motors shall be sized not smaller than _____ percent of the load.

    a. 100
    b. 115
    c. 125
    d. 150

26. Calculate the power consumed by an incandescent lamp having an impedance of 72 ohms and operating at 120 volts.

    a. 10 watts
    b. 60 watts
    c. 100 watts
    d. 200 watts

27. In dwelling units, hallways of _____ or more in length shall have at least one receptacle outlet.

    a. 5 feet
    b. 10 feet
    c. 15 feet
    d. 20 feet

28. A lighting outlet is installed to serve an interior stairway having more than five risers between floors and serving three floor levels of a single dwelling unit. How many listed wall-mounted control devices are required at each floor level?

    a. 1
    b. 2
    c. 3
    d. 4

29. Determine the ambient temperature of six size 12 AWG copper current-carrying conductors, type THHW, installed in a raceway within ⅞ inches of a roof with an ambient temperature of 38 °C.

    a. 10 amperes
    b. 12 amperes
    c. 15 amperes
    d. 20 amperes

30. The walls and roofs of transformer vaults, as well as the floor of any vault constructed with a vacant space or other stories below it, shall have a minimum fire resistance of _____.

    a. 1 hour
    b. 2 hours
    c. 3 hours
    d. 5 hours

31. Outlet boxes mounted in the ceilings of habitable rooms in dwelling occupancies at a location acceptable for the installation of a ceiling-suspended (paddle) fan shall comply with which of the following?

    a. Listed for the sole support of ceiling-suspended (paddle) fans
    b. An outlet box complying with the applicable requirements of 314.27 and providing access to structural framing capable of supporting a ceiling-suspended (paddle) fan bracket or equivalent
    c. Either a or b
    d. Both a and b

32. What size ventilated cable tray would be required for the installation of three 4/0 AWG multiconductor cables having an outside diameter of 2 inches and eight 6 AWG multiconductor cables having an outside diameter of ¾ inches?

   a. 8-inch
   b. 9-inch
   c. 12-inch
   d. 16-inch

33. Underground service conductors shall be installed in any of the following wiring methods, EXCEPT:

   a. Rigid metal conduit (RMC)
   b. Intermediate metal conduit (IMC)
   c. Electrical metallic tubing (EMT)
   d. Rigid polyvinyl chloride conduit (PVC)

34. Calculate the minimum conductor ampacity required for conductors from the output terminals of a 22-kilowatt, single-phase, 120/240-volt generator having a nameplate current rating of 92 amperes to the first distribution device containing overcurrent protection. The generator is not designed to prevent overloading.

   a. 92 amperes
   b. 100 amperes
   c. 105.8 amperes
   d. 115 amperes

35. How many outdoor receptacles are required for a one-family dwelling consisting of both a front and back entrance, heating and air conditioning equipment located on the side of the dwelling unit more than 25 feet from the front and back entrance receptacles, and a separate entryway for a porch located away from the main entrances?

   a. 1
   b. 2
   c. 3
   d. 4

36. For indoor installations, the building or structure grounding electrode system shall be used as the _____ for the separately derived system.

   a. Grounding electrode
   b. Grounding electrode conductor
   c. Grounded conductor
   d. Grounding conductor

37. In health care facilities, the cover plates for the electrical receptacles supplied from the _____ shall have a distinctive color or marking so as to be readily identifiable.

    a. Life safety branch
    b. Critical branch
    c. Normal branch
    d. Both a and b

38. A pull box is installed to contain splices on insulated circuit conductors that are 4 AWG or larger and required to be insulated. Two rows of conduits enter the box on the same wall. One row of conduits consists of four 1-inch EMT conduits and two 2-inch EMT conduits. The other row of conduits consists of three 2-inch in conduits. Determine the minimum distance between the conduit entry inside the box and the opposite wall of the box.

    a. 12 inches
    b. 16 inches
    c. 18 inches
    d. 24 inches

39. Dry-type transformers, 1,000 volts nominal or less, located in the open on walls, columns, or structures, _____.

    a. Shall be of the totally enclosed type
    b. Shall be of the non-ventilated type
    c. Shall be required to be readily accessible
    d. Shall not be required to be readily accessible

40. How may lighting outlets are required for a dwelling containing three outdoor entrances and two garage doors?

    a. 1
    b. 3
    c. 4
    d. 5

41. Receptacles installed in bathrooms shall be located within _____ of the outside edge of each sink.

    a. 1 foot
    b. 2 feet
    c. 3 feet
    d. 4 feet

42. A branch circuit supplies a combination of continuous and non-continuous loads. The continuous load is 6 amperes and the non-continuous load is 7 amperes. What is the minimum required standard ampere rating for this branch circuit?

    a. 15 amperes
    b. 20 amperes
    c. 25 amperes
    d. 30 amperes

43. Determine the number of wall space receptacles required to be installed in the room shown in the figure below.

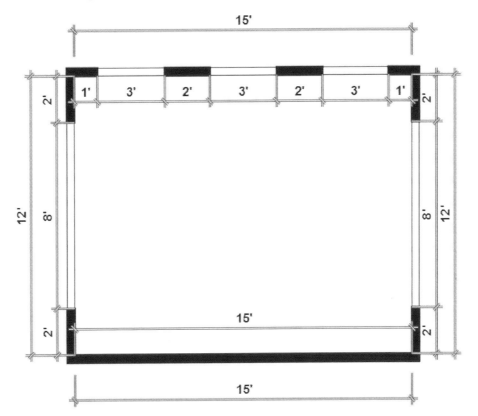

    a. 4
    b. 5
    c. 6
    d. 7

44. For Type IMC and RMC, where a conduit enters a ____, a bushing shall be provided to protect the wires from abrasion unless the ____ is designed to provide such protection.

    a. Box
    b. Fitting
    c. Enclosure
    d. Any of the above

45. The following symbol found on a construction plan would most likely indicate what electrical device?

a. Duplex receptacle
b. Recessed luminaire
c. Recessed receptacle
d. Range receptacle

46. An energy storage system for one- and two-family dwelling units shall not exceed _____ volts DC between conductors or to ground.

a. 100
b. 120
c. 150
d. 240

47. For countertop receptacle outlets in kitchens, dining rooms, etc., receptacles shall be installed so that no point along the wall line is more than _____ measured horizontally from a receptacle outlet in that space.

a. 1 foot
b. 2 feet
c. 3 feet
d. 4 feet

48. Determine the smallest allowable insulated conductor for the secondary of field-installed neon lighting having a voltage of 1,000 volts or less nominal.

a. 20 AWG
b. 18 AWG
c. 16 AWG
d. 14 AWG

49. Determine the minimum size grounded conductor of an overhead service whose ungrounded conductors are 2/0 AWG copper and terminate at a 200-ampere breaker.

a. 2 AWG copper or 1/0 AWG aluminum
b. 4 AWG copper or 2 AWG aluminum
c. 6 AWG copper or 4 AWG aluminum
d. 8 AWG copper or 6 AWG aluminum

50. Six feet of 1¼-inch liquid-tight flexible metal conduit (LFMC) is used to connect equipment where flexibility is required for the equipment to be moved after installation. Which of the following conditions would allow the LFMC to be used as an equipment grounding conductor?

a. The conduit is terminated in listed fittings.
b. There are 8 feet of flexible metal conduit (FMC) in the effective ground-fault path.
c. The circuit conductors are protected by an overcurrent device rated 80 amperes.
d. The use of LFMC as an equipment grounding conductor is prohibited for this installation. A wire-type EGC is required.

51. Which of the following can be used as a source of electrical energy?

a. Magnetism
b. Pressure
c. Friction
d. All of the above

52. If required by the authority having jurisdiction, a diagram showing feeder details shall be provided prior to the installation of the feeders. Such a diagram shall show _____.

a. The area in square feet of the building or other structure supplied by each feeder
b. The total calculated load before applying demand factors, the demand factors used, and the calculated load after applying demand factors
c. The size and type of conductor to be used
d. All of the above

53. In dwelling units, receptacles shall be installed such that no point measured horizontally along the floor line of any wall space is more than _____ from a receptacle outlet.

a. 5 feet
b. 6 feet
c. 10 feet
d. 12 feet

54. What is the minimum ampacity rating allowed for a three-wire service on a one-family dwelling?

a. 70 amperes
b. 100 amperes
c. 125 amperes
d. 200 amperes

55. At least one receptacle must be installed for countertop and work surfaces that are _____ or wider in kitchens, pantries, breakfast rooms, dining rooms, and similar areas of dwelling units.

 a. 12 inches
 b. 18 inches
 c. 24 inches
 d. 30 inches

56. Where a monorail hoist or hand-propelled crane bridge installation meets which of the following, the runway contact conductors disconnecting means shall be permitted to be omitted?

 a. The unit is controlled from the ground or floor level.
 b. The unit is within view of the power supply disconnecting means.
 c. No fixed work platform has been provided for servicing the unit.
 d. All of the above are required.

57. Calculate the required service size for a multifamily dwelling consisting of 10 units. Each unit contains a single-phase 120/208 -volt panel and is fed from the three-phase 120/208-volt service. The laundry facility is located elsewhere on the premises. Each unit is 1,200 square feet, and includes a 12-kW range, a 5-kW water heater, a 1,200 VA dishwasher, and a 900 VA disposal. Each unit has a 2-ton air conditioning unit (2,880 VA) and a 5-kW heating unit.

 a. 500 amperes
 b. 450 amperes
 c. 400 amperes
 d. 350 amperes

58. Three 6 AWG THWN copper motor feeder conductors supply power to one 20-horsepower, three-phase, 460-volt induction motor; and two 10 horsepower, three-phase, 460-volt induction motors. The branch-circuit short-circuit and ground-fault protection for the motors are inverse time breakers. Determine the maximum rating of the inverse time breaker used for motor feeder short-circuit and ground-fault protection.

 a. 60 amperes
 b. 80 amperes
 c. 90 amperes
 d. 100 amperes

59. All receptacles rated 125 volts through 250 volts, 60 amperes or less, located within _____ of the inside walls of a pool shall have GFCI protection or SPGFCI protection as required.

 a. 10 feet
 b. 15 feet
 c. 20 feet
 d. 25 feet

60. A 240-volt fixed electric space heater has 14.5 kW of resistive heating elements, one 11-kW element and one 3.5-kW element. What is the rating(s) of the overcurrent protection device(s) required to supply this equipment?

    a. 60 amperes
    b. 80 amperes
    c. 50 amperes and 15 amperes
    d. 60 amperes and 20 amperes

61. What is the minimum conductor ampacity for a branch-circuit that supplies three arc welders? The I_{1eff} of the first two arc welders is 34 amperes. The rated primary current of the third welder is 23 amperes and is a nonmotor generated welder with a rated duty cycle of 60%.

    a. 83.25 amperes
    b. 85.94 amperes
    c. 91 amperes
    d. 100 amperes

62. Determine the smallest allowable service conductors for a service with five disconnects having a combined overcurrent rating of 230 amperes and a calculated load of 185 amperes.

    a. 1/0 AWG
    b. 2/0 AWG
    c. 3/0 AWG
    d. 4/0 AWG

63. In general, Type AC cable shall be secured within _____ of every outlet box, junction box, cabinet, or fitting; and at intervals not exceeding _____.

    a. 12 inches, 4½ feet
    b. 36 inches, 6 feet
    c. 12 inches, 6 feet
    d. 36 inches, 4½ feet

64. On a four-wire, delta-connected service where the midpoint of one phase winding is grounded, the service conductors having the higher phase voltage to ground shall be durably and permanently marked by an outer finish that is _____ in color, or by other effective means, at each termination or junction point.

    a. Orange
    b. Purple
    c. Red
    d. Yellow

65. An enclosure that does not contain a device(s) other than splicing devices or supports a luminaire(s), a lampholder, or other equipment and is supported by entering raceways shall not exceed _____ in size. It shall have threaded entries or identified hubs. It shall be supported by two or more conduits threaded wrench-tight into the enclosure or hubs.

    a. 75 cubic inches
    b. 100 cubic inches
    c. 200 cubic inches
    d. 300 cubic inches[3]

66. Disconnecting means for air-conditioning and refrigerating equipment shall be located within _____ and be _____ from the air-conditioning or refrigerating equipment.

    a. 25 feet, accessible
    b. Sight, readily accessible
    c. 50 feet, readily accessible
    d. Sight, accessible

67. Incandescent lamp luminaires shall be marked to indicate the maximum allowable wattage of lamps. The marking shall be permanently installed, in letters at least _____ high, and shall be located where visible during relamping.

    a. ¼ inch
    b. ½ inch
    c. ¾ inches
    d. 1 inch

68. A 45-kVA dry-type transformer serves as the source of a separately derived system. What is the size of the wire-type supply-side bonding jumper required to bond the transformer to the enclosure containing the first means of disconnect if the ungrounded conductors are 1 AWG copper?

    a. 8 AWG copper
    b. 6 AWG copper
    c. 4 AWG copper
    d. 2 AWG copper

69. Electrically operated in-sink disposers shall be permitted to be cord-and-plug connected with a flexible cord identified as suitable in the installation instructions of the appliance manufacturer where all of the following conditions are met, EXCEPT:

    a. The length of the cord is not less than 18 inches and not exceeding 36 inches.
    b. The receptacle is accessible and located to protect against physical damage to the flexible cord.
    c. The flexible cord has an equipment grounding conductor and is terminated with a grounding-type attachment plug.
    d. The receptacle is of the self-grounding type.

70. Which of the following is NOT a prohibited use of Type NM and NMC cables?

a. Exposed within a dropped or suspended ceiling cavity in any building other than one- and two-family and multifamily dwellings.
b. As service-entrance cable.
c. Types I and II construction, when installed in raceways permitted for Types I and II construction.
d. In motion picture studios.

71. Underground service entrance conductors, where subject to physical damage, shall be protected by any of the following means, EXCEPT:

a. Rigid metal conduit (RMC).
b. Electrical metallic tubing (EMT).
c. Schedule 80 polyvinyl chloride conduit (PVC).
d. Schedule 40 polyvinyl chloride conduit (PVC).

72. Determine the length of a straight pull box where two 1-inch rigid metal conduits (RMC) and one 2-inch rigid metal conduit (RMC) enter a box on one side. A single 4-inch rigid metal conduit enters the box on the opposite side. The largest conductor contained in the raceway is 2/0 THW.

a. 16 inches
b. 24 inches
c. 30 inches
d. 32 inches

73. In hotel, motel, or similar occupancies, guest rooms or guest suites shall have at least one lighting outlet installed in every habitable room or bathroom controlled by any of the following, EXCEPT:

a. A listed wall-mounted control device.
b. A pull chain fixture, provided the lighting outlet is not controlled from more than one location.
c. Outside of bathrooms and kitchens, one or more receptacles controlled by a listed wall-mounted control device.
d. An occupancy sensor in addition to a listed wall-mounted control device or an occupancy sensor located at a customary wall switch location and equipped with a manual override that allows the sensor to function as a wall switch.

74. The disconnecting means for motor circuits rated 1,000 volts nominal or less shall have an ampere rating not less than _____ of the full-load current rating of the motor as determined by_____.

a. 115%; Table 430.247, Table 430.248, Table 430.249, or Table 430.250
b. 115%; the nameplate full-load current of the motor
c. 125%; Table 430.247, Table 430.248, Table 430.249, or Table 430.250
d. 125%; the nameplate full-load current of the motor

75. Determine the maximum standard ampere rating for the overcurrent protection devices of both the primary and the secondary of a three-phase 60 kVA transformer. The primary voltage is three-phase, 480 volts and the secondary voltage is three-phase, 120/208-volt wye.

    a. Primary = 200 A, secondary = 225 A
    b. Primary = 200 A, secondary = 200 A
    c. Primary = 225 A, secondary = 200 A
    d. Primary = 180 A, secondary = 208 A

76. A large industrial facility consisting of large inductive motor loads is being charged a power factor surcharge by the utility provider due to the facility's poor power factor. What is the best course of action for correcting the poor power factor?

    a. Upgrade older high-intensity discharge lighting to newer LED light fixtures
    b. Install power factor correction capacitors at large induction motors
    c. Increase the size of the service entrance conductors
    d. Install variable frequency drives (VFDs) on any induction motors that are currently operating near their full load current

77. Where a continuous industrial process experiencing a nonorderly shutdown would introduce additional or increased hazards, ground-fault protection of equipment is required on each feeder disconnect rated _____ amperes or more and installed on solidly grounded wye electrical systems of more than _____ volts to ground, but not exceeding 1000 volts phase-to-phase.

    a. 400, 120
    b. 400, 150
    c. 1,000, 150
    d. Ground-fault protection of equipment is not required on this installation.

78. A transformer has 1,200 turns on the primary winding and 300 turns on the secondary winding. If there are 480 volts AC across the primary winding, how many volts will be induced on the secondary winding?

    a. 480 volts
    b. 240 volts
    c. 120 volts
    d. 1,920 volts

79. Considering 230.79 and 230.80, what is the minimum combined ampacity rating for a one-family dwelling having more than one service disconnect and a calculated load of 80 amperes?

    a. 80 amperes
    b. 100 amperes
    c. 125 amperes
    d. 150 amperes

80. For feeders supplied form direct-current systems, the positive polarity on conductors size 6 AWG or smaller shall be identified by any of the following means, EXCEPT:
   a. Red marking tape.
   b. A continuous red outer finish.
   c. A continuous red stripe, durably marked along the conductor's entire length on insulation other than green, white, gray, or black.
   d. Imprinted plus signs (+) or the word "Positive" or "POS" durably marked on insulation of a color other than green, white, gray, or black and repeated at intervals not exceeding 24 inches.

81. Outlet boxes that do not enclose devices or utilization equipment shall have a minimum internal depth of _____.
   a. ½ inch
   b. ¾ inches
   c. 1 inch
   d. 2 inches

82. Determine the number of 12 AWG conductors that can be pulled through a 4-inch by 1½-inch box if the box already contains one receptacle and two 12 AWG conductors, as well as one 12 AWG equipment grounding conductor, if the length of the conductors pulled through the box is less than 12 inches.
   a. 4
   b. 5
   c. 6
   d. 7

83. Determine the minimum rating of a disconnecting means for a phase converter supplying a three-phase motor with a full load amperage of 60 amperes, when the disconnecting means is a molded-case switch.
   a. 100 amperes
   b. 125 amperes
   c. 150 amperes
   d. 200 amperes

84. In theaters, television studios, and similar locations, footlights, border lights, and proscenium sidelights shall be arranged so that no branch circuit supplying such equipment carries a load exceeding _____.
   a. 15 amperes
   b. 20 amperes
   c. 25 amperes
   d. 30 amperes

85. Determine the required volume for a box that will contain two receptacles, two 12 AWG conductors, two 10 AWG conductors, and one 10 AWG equipment grounding conductor. The box will incorporate an internal cable clamp. One receptacle will be connected to the 12 AWG conductors and one receptacle will be connected to the 10 AWG conductors.

a. 19 cubic inches
b. 21.5 cubic inches
c. 24 cubic inches
d. 25.5 cubic inches

86. Receptacle outlets installed in dwelling units for specific equipment, such as laundry equipment, shall be installed within _____ of the intended location of the appliance.

a. 3 feet
b. 4 feet
c. 5 feet
d. 6 feet

87. Determine the rating of an inverse time breaker installed as the motor branch-circuit short-circuit and ground-fault protection for a 20-horsepower, three-phase, 230-volt induction motor.

a. 110 amperes
b. 125 amperes
c. 150 amperes
d. 175 amperes

88. Determine the ampacity of 15 size 14 AWG copper type XHHW-2 current-carrying conductors installed in a raceway with an ambient temperature of 40 °C.

a. 11.38 amperes
b. 15.92 amperes
c. 18.48 amperes
d. 30.27 amperes

89. Under what conditions are distribution blocks allowed to be installed in pull and junction boxes less than 100 cubic inches in volume?

a. When installed with an appropriate insulating barrier
b. When the distribution block is listed for use in smaller enclosures
c. When the enclosure is listed for use with distribution blocks
d. When the distribution block is used for connection of equipment grounding conductors

90. For generators larger than 15 kilowatts, what must be provided for remote emergency shutdown of the prime mover?

    a. A stop switch located outside the equipment room or generator enclosure

    b. Provisions to disable all prime mover start control circuits, rendering the prime mover incapable of starting

    c. Initiate a shutdown mechanism that requires a mechanical reset

    d. All of the above

91. Cartridge fuses in circuits of any voltage, and all fuses in circuits over _____ volts to ground, shall be provided with a disconnecting means on their supply side so that each circuit containing fuses can be independently disconnected from the source of power.

    a. 120

    b. 150

    c. 208

    d. 230

92. For non-metallic sheathed cable, the maximum length between the cable entry and the closest cable support shall not exceed _____.

    a. 12 inches

    b. 18 inches

    c. 20 inches

    d. 24 inches

93. Which of the following is not a special requirement for large-scale photovoltaic electric supply stations?

    a. The use of medium- or high-voltage switch gear, substation, switch yard, or similar methods to interconnect the PV system safely and effectively to the utility.

    b. The electrical loads within the PV electric supply station shall only be used to power auxiliary equipment for the generation of PV power.

    c. Large-scale PV electric supply stations shall not be installed on buildings.

    d. Large-scale PV electric supply stations shall not exceed 50 kV to ground.

94. Where an outlet branch-circuit-type AFCI is installed at the first outlet, what additional installation requirements are necessary?

    a. The combination of AFCI outlet and branch circuit overcurrent protection shall be listed to provide system combination-type arc fault protection in addition to continuous "home run" conductors of no more than 50 feet for 14 AWG and 70 feet for 12 AWG. The first outlet marked to indicate it is the first outlet of the circuit.

    b. A metal raceway, metal wireway, metal auxiliary gutter, or Type MC or AC cable are installed for the portion of the branch circuit between the branch-circuit overcurrent device and first outlet.

    c. Listed metal or non-metallic conduit or Type MC cable encased in no less than 2 inches of concrete for the portion of the branch circuit between the branch-circuit overcurrent device and the first AFCI outlet to provide protection for the remaining circuit.

    d. Any of the above.

95. When installing receptacle outlets to serve an island or peninsular countertop or work surface, which of the following is NOT an allowable receptacle outlet location?

    a. Below countertop or work surfaces, when not more than 12 inches below the countertop or work surface

    b. On or above, but not more than 20 inches above, a countertop or work surface

    c. In a countertop using receptacle outlet assemblies listed for use in countertops

    d. In a work surface using receptacle outlet assemblies listed for use in work surfaces or listed for use in countertops

96. Boxes required to support ceiling-mounted luminaires weighing more than _____ shall be supported independently of the outlet box, unless the outlet box is listed for not less than the weight supported.

    a. 25 pounds

    b. 35 pounds

    c. 50 pounds

    d. 70 pounds

97. Ground-fault protection of equipment shall be provided for solidly grounded wye electric service of more than _____ volts to ground but not exceeding 1,000 volts phase-to-phase for each service disconnect rated _____ amperes or more.

    a. 120, 400

    b. 150, 400

    c. 120, 1,000

    d. 150, 1,000

98. Determine the minimum distance between raceway entries enclosing the same 2/0 AWG THHW circuit conductors entering a pull box on adjacent walls. Both conduits are 2-inch EMT.

    a. 12 inches
    b. 14 inches
    c. 16 inches
    d. 20 inches

99. Determine the number of 125-volt, 15- or 20-ampere receptacle(s) required for a show window measuring 15 linear feet measured horizontally at its maximum width.

    a. 1
    b. 2
    c. 3
    d. 4

100. Type MC cable shall be permitted to be unsupported and unsecured where the cable complies with which of the following?

    a. Is fished between access points through concealed spaces in finished buildings and supporting is impractical.
    b. Is not more than 6 feet in length from the last point of cable support to the point of connection to luminaires or other equipment in an accessible ceiling.
    c. Is Type MC of the interlocked armor type in lengths not exceeding 3 feet from the last point where it's securely fastened and used to connect equipment where flexibility is necessary.
    d. Any of the above

Answer Key and Explanations

1. C: Article 100. New to the 2023 NEC is the definition of special-purpose GFCI devices. They are intended for higher voltage circuits where regular GFCI may not provide enough protection, such as pool house equipment rooms.

2. B: 245.21(C)(1). Distribution cutout above 1,000 volts nominal shall not be used indoors, underground, or in metal enclosures. They must also be located so that they may be readily and safely operated and re-fused, and so that the exhaust of the fuses does not endanger persons. Fused cutouts subject to a high energy fault often fail violently. The cutout tube containing the fuse link can act as a bore, directing the burning fuse link exhaust out both ends. Care must be taken to ensure these are not directed at locations likely to endanger persons.

3. D: 495.24. For equipment operating at over 1,000 volts ac or 1,500 volts dc, nominal, in field-fabricated installations, there must be an air gap between bare live conductors of different phases and between any of those conductors and adjacent grounded surfaces. The answer is from Table 495.24, using the value for 23 kV equipment, phase-to-ground, outdoors.

4. D: 680.9(A). Article 680 modifies the requirements for overhead clearance for service conductors found in article 230. For a conductor with a voltage of less than 750 volts to ground, the clearance in any direction to the water level, edge of water surface, base of diving platform, or permanently anchored raft must be at least 22.5 feet as found in Table 680.9(A).

5. D: 210.52(C)(2). There is no longer a requirement to install receptacle outlets to serve island or peninsular countertops. However, if receptacle outlets are not installed, provision must still be made for the future addition of such receptacles.

6. D: 215.4(B). When installing feeder conductors for AC systems in metallic raceways, all conductors, including the neutral conductor, must be grouped together to avoid inductive heating of the surrounding metal. When feeder conductors are run in parallel, one conductor from each phase, including the neutral, must be run in each raceway.

7. C: Table 450.3(B). When sizing overcurrent protection for transformers 1,000 volts and less, Table 450.3(B) is used. The first step is to determine the transformers rated current for the primary and, if necessary, the secondary of the transformer. Next, apply the percentages based on the protection method and the appropriate column for the rated current.

Rated Current of Primary

$$\frac{60,000 \text{ VA}}{(480 \text{ V})(1.73)} = 72.29 \text{ A}$$

For transformers protected on the primary as well as the secondary and having primary currents of more than 9 amperes, the percentage multiplier is 250%.

72.29 A × 2.5 = 180.73 A

Note 1 of Table 450.3(B) states that when this current does not correspond to a standard rating of a fuse or nonadjustable circuit breaker, a higher rating that does not exceed the next higher standard rating shall be permitted. Standard overcurrent ratings can be found in 240.6(A).

180.73 amperes or the next highest standard rating of 200 amperes.

Rated Current of Secondary

$$\frac{60,000 \text{ VA}}{(208 \text{ V})(1.73)} = 166.74 \text{ A}$$

For transformers protected on the primary as well as the secondary and having secondary currents of more than 9 amperes, the percentage multiplier is 125%.

166.74 A × 1.25 = 208.43 A

Note 1 of Table 450.3(B) states that when this current does not correspond to a standard rating of a fuse or nonadjustable circuit breaker, a higher rating that does not exceed the next higher standard rating shall be permitted. Standard overcurrent ratings can be found in 240.6(A).

208.43 amperes or the next highest standard rating of 225 amperes.

8. B: 430.6(A)(1), Table 430.248, Table 430.52(C)(1). When sizing motor branch-circuit short-circuit and ground-fault protection for single-phase AC motors, Table 230.248 is used to determine the current that will be applied to Table 430.52(C)(1) based on the horsepower and rated voltage of the motor. The current listed for a 7½ horsepower, 208-volt, single-phase motor is 44 amperes.

Table 430.52(C)(1) gives the maximum rating or setting for the motor branch-circuit short-circuit and ground-fault protection as a percentage of the current from Table 430.248.

44 A × 1.75 = 77 A

Exception No. 1 Table 430.52(C)(1) allows the next highest standard ampere rating for a protection device to be used when the calculated rating does not correspond to a standard ampere rating. In this case, 77 amperes do not correspond to a standard ampere rating, and the next highest standard rating is 80 amperes.

Note: *Motor branch-circuit short-circuit and ground-fault protection* must be sized large enough to prevent nuisance tripping during startup due to inrush current, but still protect against line-to-line and line-to-ground faults within the motor circuit and motor. *Motor overload protection* protects against overloads above the designed limitations of the motor circuit conductors and motor, but below the ratings of the *branch-circuit short-circuit and ground-fault protection.*

9. B: 700.12(C)(4). Emergency power systems generally must be of suitable rating and capacity to supply and maintain the total load for not less than 2 hours, but in the case that they are used for emergency illumination, that requirement drops to 1½ hours.

10. B: 314.28(A)(1). Pull boxes containing conductors larger than 4 AWG and required to be insulated must be sized according to 314.28(A). For straight pulls, the length of the box must be at least eight times the diameter of the largest raceway.

3 inches × 8 = 24 inches

11. C: 517.32, 517.35(A). Many equipment branches are required to be automatically connected to an alternate source of power. The branches in 517.35(A) shall be permitted to do so with a delay. Exit signs must be arranged to have an automatic, non-delayed connection to an alternate source of power. This is most commonly done through an internal battery.

12. B: 694.23(A). Unless a wind turbine has a swept area less than 538 square feet, it shall have a readily accessible shutdown switch or button that results in a parked turbine state. The shutdown procedure for a wind turbine shall be defined and permanently posted at the location of shutdown means and at the location of the turbine controller or disconnect, if the location is different, regardless of the swept area of the turbine.

13. C: 620.24(A). Elevator hoistway pit lighting shall not be connected to the load side of a ground-fault circuit interrupter. Elevator hoistway pits are both dark and frequently subject to ground water intrusion. Lighting should be installed above the point where it might be subject to standing water in the event a sump pump is not functioning properly, and lighting should not be connected to the secondary of a GFCI, leading to darkness in the elevator pit with no way to reset it.

14. B: No point measured horizontally along a wall space should be more than 6 feet from a receptacle outlet, and any wall space 2 feet or more in width requires a receptacle. Therefore, five receptacles are required for this room.

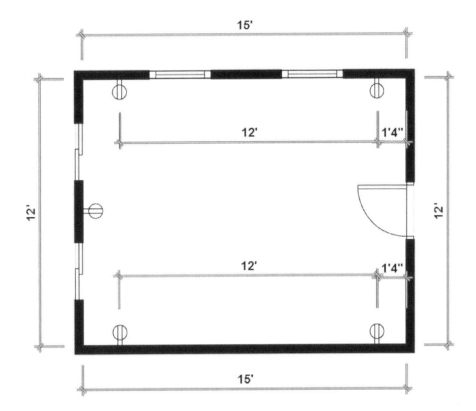

15. A: 702.11(B). When a portable generator is used as the source of an optional standby system and is used as a separately derived system, it shall be grounded to the grounding electrode in accordance with 250.30. When the portable generator is used as a non-separately derived system, the equipment grounding conductor shall be bonded to the system grounding electrode. By definition, a separately derived system is an electrical source, other than a service, having no direct connection(s) to circuit conductors of any additional electrical source other than those established by grounding and bonding connections. In practice, the usual determining factor will be whether or not the service and the alternate source of power share a common, unswitched grounded conductor. If the grounded (neutral) conductor is unswitched, the alternate source of power is considered a non-separately derived system.

16. B: Table 352.30(B). Rigid PVC is more flexible at smaller sizes and thus requires closer spacing between supports than that of larger rigid PVC conduits.

17. C: 240.24. The maximum height above the floor or the working platform that a switch or circuit breaker is allowed to be installed is 6 feet 7 inches. This is measured from the center of the grip of the operating handle when in its highest position.

18. A: 460.10. Capacitors may not be encountered as frequently as many other electrical devices. However, they are frequently used for starting and running single-phase motors, surge protection devices, and power factor correction. The case with the most capacitors must be connected to the equipment grounding conductor to allow the overcurrent protection device to trip if the capacitor shorts to its case.

19. C: 110.22(A). Unless the disconnecting means is arranged and located so as to make its purpose evident, each disconnecting means must be legibly and durably marked to indicate its purpose. In all applications other than one- and two-family dwellings, the marking must include the identification for the circuit source that supplies the disconnecting means, i.e., a panel and circuit number. While industrial settings are obviously not considered one- or two-family dwellings, they do not adequately cover all areas where such identification is required.

20. D: 230.24(B) provides the required vertical clearances for overhead service conductors. Commercial driveways subject to truck traffic require that the vertical clearance meet the requirements of 230.24(B)(4) or 18 feet.

21. A: 210. 52(A)(2). For the purpose of wall receptacle placement, any wall space 2 feet or more in width requires a receptacle. These measurements must consider connecting wall spaces, even around corners, so long as they are unbroken along the floor line.

22. D: 210.52(A)(3). Floor receptacles installed within 18 inches of a wall are permitted to take the place of a wall-mounted receptacle.

23. A: 342.24(B). As is common with most conduit type raceways, the maximum amount of bend in a conduit run between pull boxes is 360 degrees. It is good practice to provide a pull box at 100-foot intervals where practical. However, contrary to common understanding, this is not an NEC requirement.

24. A: 525.21. In addition to the requirement to be within sight as well as within 6 feet of the operator's station, if the area is accessible to unqualified persons, the disconnecting means shall be lockable. A shut trip device that opens the fused disconnect or circuit breaker when a switch located in the ride operator's console is closed shall be a permissible method of opening the circuit.

25. C: 424.4(B). Though the operation of fixed electric space heaters may be intermittent, there may be times when the equipment runs for extended periods of time. The NEC requires that fixed electric space heating be treated as a continuous load and sized accordingly.

26. D: Ohm's law states that electrical current is directly proportional to voltage and inversely proportional to resistance. Additionally, electrical power is directly proportional to both current and voltage, and inversely proportional to resistance. Ohm's law can be expressed with the following two formulas.

$$E = I \times R$$

and

$$P = I^2R \text{ or } P = \frac{E^2}{R} \text{ or } P = I \times E$$

E = Voltage or *electromotive force*

I = Current or *intensity*

R = Resistance

P = Power

Another term to become familiar with is impedance. Impedance is the effective resistance of an electric circuit or component to alternating current, arising from the combined effects of ohmic resistance and reactance. Impedance is analogous to resistance, but considers all oppositions to current flow, such as inductive reactance and capacitive reactance, both of which are common in alternating current circuits.

To solve this problem, simply choose the equation to solve for power that uses the known values of voltage and resistance. For this equation, simply substitute resistance with impedance (Z).

$$P = \frac{E^2}{Z}$$
$$P = \frac{120^2}{72}$$
$$P = \frac{14,400}{72}$$
$$P = 200 \text{ watts}$$

27. B: 210.52(H). At least one receptacle is required in each hallway 10 feet or more in length. Additional requirements exist for foyers that are not part of the hallways referenced in 210.52(H). Foyers having an area greater than 60 square feet shall have a receptacle(s) located in each wall space 3 feet or more in width.

28. C: 210.70(A)(2)(3) requires that a listed wall-mounted control device be installed at each floor landing of a stairway to control the stairway lighting outlets. This is to allow for the ability to control the lighting regardless of which floor the occupant is on.

29. B: Table 310.16 is used for determining the ampacity of insulated conductors with not more than three current-carrying conductors in a raceway, cable, or directly buried. Table 310.16 is also based on an ambient temperature of 30 °C (86 °F). When these conditions change, electricians must adjust the ampacity of the conductors in accordance with 310.15(B) and (C). Temperature corrections and adjustment factors shall be permitted to be applied to the ampacity for the temperature rating of the conductor. Though electricians often base conductor sizing off of the temperature limitations of terminations and equipment, when adjusting for temperature or conductor bundling, they are permitted to do the adjustment based on the conductor temperature rating, i.e., the 90 °C column for THHN.

The conditions of use for the conductors in this question require an adjustment of the ampacity to compensate for both the increased ambient temperature beyond the design of Table 310.16, as well as the adjustment required for more than three current-carrying conductors. Table 310.15(B)(1) is used to determine the ambient temperature adjustment factor when the ambient temperature is not 30 °C. With an ambient temperature of 40 °C, the ampacity of the conductors must be adjusted by a factor of 0.91. Next, Table 310.15(C)(1) is used to determine the adjustment factor when more than three current-carrying conductors are installed together. Using Table 310.15(C)(1) reveals that an adjustment factor of 50% must be applied to the ampacity of the conductors. Based on Table 310.16, the ampacity of 14 AWG copper XHHW-2 is 25 amperes.

$$25 \text{ A} \times 0.91 = 22.75 \text{ A}$$

$$22.75 \text{ A} \times 0.5 = 11.38 \text{ A}$$

30. C: 450.42. The walls and roofs of transformer vaults must be constructed with a minimum of three hours of fire resistance. Additionally, for transformer vault floors in contact with the earth, they shall be constructed of concrete that is no less than 4 inches thick. Vault floors where the vault is constructed with vacant space or other stories below must also have a minimum of three hours of fire resistance. The exception to 450.42 allows for a 1-hour rating where transformers are protected with an automatic sprinkler, water spray, carbon dioxide, or halon systems.

31. C: 314.27(C). The requirements of 314.27(C) apply when a fan is either installed initially or could be installed in the future. These boxes must be either listed for the sole support of ceiling-suspended (paddle) fans or must comply with the applicable requirements of 314.27(C).

32. C: 392.22(A)(1)(c). When cables of 4/0 AWG or larger are installed in the same cable tray with cables having conductors smaller than 4/0 AWG, an electrician must calculate the allowable fill based on Table 392.22(A), Column 2 for ventilated tray and Column 4 for solid tray. The term "Sd" in Columns 2 and 4 is equal to the sum of the diameters of all cables 4/0 AWG and larger.

Three cables with 2-inch outside diameter = 6 inches

1.2 × Sd

1.2 × 6 = 7.2

Now to add the cross-sectional area of the cables smaller than 4/0 AWG

$$A = \pi r^2$$

$$A = 3.14(0.375^2)$$

$A = 0.44$ square inches

Total area of all cables = 0.44 square inches $\times$ 8 = 3.53 square inches

Total of cables 4/0 AWG and larger plus cables smaller than 4/0 AWG =

7.2 square inches + 3.53 square inches = 10.73 square inches

In this instance, a cable tray that has an allowable fill of 10.73 square inches – 1.2 (Sd) or larger, or a 12-inch cable tray would be needed.

33. C: Section 230.30(B) lists all the acceptable wiring methods for service conductors installed underground. Of the wiring methods listed, electrical metallic tubing (EMT) is not present. Type EMT conduit does not have the inherent corrosion resistance of other wiring methods and is a poor choice for any underground application.

34. C: 445.13(A). For generators that are not designed to prevent overloading, the ampacity of the conductors between the output terminals of the generator and the first overcurrent protection device must have an ampacity of at least 115% of the nameplate current rating of the generator.

$92 \times 1.15 = 105.8$

35. C: Several locations require outdoor receptacles in dwelling units. At least two outdoor receptacles will be required for every dwelling unit. 210.52(E)(1) requires that at least one receptacle be installed in both the front and back of the dwelling. 210.63 and 210.63(A) require that an outdoor receptacle be installed within 25 feet of heating, air-conditioning, and refrigeration equipment. Finally, at least one receptacle must be located within the perimeter of any porch, deck, or balcony that is attached to or located within 4 inches of the dwelling. If these receptacles can be located so that their required installation locations overlap, one such receptacle could be used to meet the requirements of multiple sections relating to required outdoor receptacles. In general, these receptacles must be located within 6½ feet of grade level or porch or balcony surface.

36. A: 250.30(A)(4). All electrodes present in an installation must be bonded together to form a grounding electrode system. Separately derived systems are no different, and bonding all electrodes to a grounding electrode system ensures that no potential difference is created between the earth and exposed, normally non-current carrying metal parts.

37. D: 517.31(E). Receptacles that are supplied by the life safety branch and the critical branch shall have a distinctive color or marking used for identification. Receptacles supplied from these branches are often required to indicate the panelboard and branch-circuit number supplying them.

38. C: 314.28(A)(2). For angle, U pull, or splice boxes, the minimum distance between the wall where the conduits enter a box and the opposite wall must be six times the diameter of the largest conduit plus the diameter of each conduit in that same row. The larger of the two minimum distances should be used. Both must be calculated:

Conduit row 1: Four 1-inch EMT and three 2-inch EMT

2 inches × 6 = 12 inches

12 inches + 2 inches + 1 inch + 1 inch + 1 inch + 1 inch = 18 inches

Conduit row 2: Two 2-inch EMT

2 inches × 6 = 12 inches

12 inches + 2 inches + 2 inches = 16 inches

The minimum distance required between the conduit entry inside the pull box and its opposite wall is then 18 inches.

39. D: 450.13(A). When dry-type transformers 1,000 volts or less are installed in the open on walls, columns, or structures, they shall not be required to be readily accessible. 450.13 requires that all transformer and transformer vaults shall be readily accessible to qualified personnel for inspection and maintenance, except for the above scenario and the scenario described in 450.13(B).

40. B: 210.70(A)(2)(2). For dwelling units, at least one lighting outlet controlled by a listed wall-mounted control device shall be installed to provide illumination on the exterior of the grade level entrance or exit. The NEC does not consider a garage door to be an entrance or exit, and therefore does not require a lighting outlet.

41. C: 210.52(D). At least one receptacle outlet must be installed in each bathroom, and it must be located within 3 feet of each sink. If more than one sink exists, a single receptacle located within 3 feet of both sinks will fulfill the requirements of this section. The receptacle must be located on a wall or partition adjacent to the sink or installed on the side or face of the sink cabinet. The receptacle must not be installed more than 12 inches below the countertop.

42. A: 210.20(A) requires that the ampacity of the branch circuit overcurrent protection device be based on 125% of the continuous load plus 100% of the non-continuous load.

Continuous load = 6 amperes × 1.25 = 7.5 amperes

Non-continuous load = 7 amperes

Total load = 14.5 amperes.

Standard ampere ratings are listed in 240.6. Based on this, 15 amperes is the minimum standard ampere rating.

43. C: Because each of the four wall segments along the top portion of the drawing is 2 feet or larger in length, they will each require a receptacle. Note that if the door openings along the top were windows, they would not be considered separate wall sections and would need to be measured as one continuous wall. The lower portion of the drawing must have receptacles within 6 feet of the openings, measured horizontally along the wall, and have no more than 12 feet between receptacles.

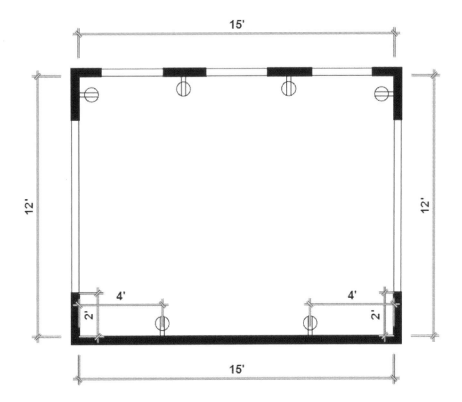

44. D: 342.46 and 344.46. For IMC, RMC, and several other wiring methods, the NEC requires that, where a conduit enters a box or other enclosure, a bushing shall be provided to protect the wires from abrasion unless the box, fitting, or enclosure is designed to provide such protection. An example of one such fitting is the Myers hub. This type of fitting has an integral plastic "throat" to protect conductors from abrasion by the threads. If this bushing requirement exists for a wiring method, the requirement can be found in section XXX.46 of the article for each wiring method. These requirements are in addition to the requirements of 300.4(G).

45. D: This symbol is most representative of a range receptacle. Symbols for receptacles contain one or more lines that transect a circle. For wall-mounted receptacles, the lines will continue past the edge of the circle and connect to the wall on which it will be mounted. Each line will represent either the number of receptacles in the enclosure or, in the case of a receptacle outlet that contains two ungrounded conductors having 240 volts between them as well as the common neutral conductor, the three transecting lines represent the number of current-carrying conductors. An example of the former is a duplex receptacle having two horizontal lines, or a quadruplex receptacle having two horizontal lines as well as two vertical lines. An example of the latter is a range or dryer receptacle such as the one used for this question. Range and dryer receptacles will most commonly be differentiated by the letter "R" or "D," respectively.

46. A: 706.20(B). Energy storage systems for one- and two-family dwellings must be designed so that the voltage between conductors or to ground does not exceed 100 volts DC. If live parts are not accessible during routine ESS maintenance, the maximum voltage between conductors or to ground can be increased to 600 volts DC.

47. B: 210.52(C)(1). Similar to the requirements for wall receptacle spacing, countertop receptacle spacing should be viewed from the perspective of the cord-and-plug connected device. Most small appliances used on countertops in kitchens are designed with a 2-foot cord. When installing countertop receptacles, the installer must ensure that a device can be placed anywhere on the counter and be within 2 feet of a receptacle. Once the installer has placed a receptacle within 2 feet of the beginning of a countertop space, spacing can be increased to 4 feet and a receptacle must be placed within 2 feet of the end of the countertop section.

48. B: 600.31(B). The smallest insulated conductor allowed for the secondary of field-installed neon lighting is 18 AWG. The contents of chapter 6, for special equipment, modify the rules of the first four chapters of the NEC. Electricians familiar with the articles within chapter 6 will be able to quickly recognize when to reference the appropriate article within that chapter for any modifications to the rules that apply generally. In this case, questions containing references to neon lighting should quickly point to Article 600.

49. B: 230.23(C) states that the minimum size grounded conductor for an overhead service must comply with 250.24(D). Among the requirements in 250.24(D), the sizing of the grounded conductor of a single raceway or cable shall not be smaller than specified in Table 250.102(C)(1). Per Table 250.102(C)(1), the size of the grounded conductor is based on the associated ungrounded conductor size. Thus, for 2/0 AWG copper ungrounded conductors, the minimum size grounded conductor is 4 AWG copper.

50. D: 350.60(A), 250.118(A)(6). Liquid-tight flexible metal conduit (LFMC) is allowed to be used as an equipment grounding conductor when the conditions meet the requirements of 250.118(A)(6). However, when LFMC is installed for the purpose of minimizing vibration from equipment or to allow movement after installation, a wire-type EGC is required to be installed.

51. D: Electrical energy can be produced through a variety of means. Electromagnetic induction is by far the most common means of generating electrical power. When a conductor passes through a magnetic field, a current is produced in the conductor as it breaks the lines of magnetic flux. Other sources of electrical energy are friction, heat, light, chemical, and pressure. Examples of devices that generate electricity in this way are Van de Graff generators, thermocouples, solar panels, batteries, and piezoelectric crystals respectively.

52. D: The NEC often allows the Authority Having Jurisdiction (AHJ)—defined by the NEC as an organization, office, or individual responsible for enforcing the requirements of a code or standard; or for approving equipment, materials, and installation or a procedure—great latitude in making the final determination as to whether or not an installation meets the requirements of the NEC. Though not common, it may be necessary to provide the AHJ with documentation detailing the relevant information used in determining certain elements of an installation. This highlights the importance of the journeyman electrician's ability to both calculate and understand engineering-level concepts such as demand load calculations. Section 215.5 is one example of the NEC providing the AHJ, the code language to withhold approval until such information is provided.

53. B: For household devices, such as televisions, floor lamps, etc., it is common to find a 6-foot cord for devices intended to be plugged into receptacles installed in wall spaces. To eliminate the use of extension cords, section 210.52(A)(1) requires that no point measured horizontally along a wall space be more than 6 feet from a receptacle outlet. Confusion can arise during the construction process if the installer interprets this code section to mean that wall receptacle outlets should be spaced 6 feet apart. In determining the location of wall receptacles during construction, the installer should adopt the point-of-view of the cord-and-plug connected device. If the device can be placed anywhere along the wall and be more than 6 feet away from a receptacle, then it is a violation. In practice, receptacles must be within 6 feet of the start of a wall, with 12-feet spacing between receptacles, and within 6 feet of the end of the wall section. These measurements are continued around corners until broken by doorways, openings, etc.

54. B: For one-family dwelling units, the minimum disconnecting means shall not be rated less than 100 amperes as per 230.79(C). For services with calculated loads above 100 amperes, the disconnecting means shall not be rated less than the calculated load.

55. A: 210. 52(C). At least one receptacle is required to be installed on countertop spaces 12 inches or wider in kitchens and other similar areas of a dwelling unit. Receptacle outlets shall not be required directly behind a range, counter-mounted cooking unit, or sink.

56. D: 610.32. For cranes and monorail hoists, the required disconnecting means is permitted to be omitted if the requirements in 610.32 are met.

57. D: The requirements for determining the service load of dwelling units using the optional method can be found in Part IV of Article 220. Care should be taken to ensure the correct calculated load is obtained. Article 220 can be somewhat complicated for those unfamiliar with the intent of the optional method. However, the steps necessary to calculate the load correctly can be distilled into just a few key ideas, many of them applying to all such calculations. For this article, it is generally accepted that kilovolt-amperes (kVA) shall be considered equivalent to kilowatts (kW).

One-family and multifamily dwelling unit service load calculations differ in the application of demand factors. In one-family dwellings, loads are separated into *general loads* and *air-conditioning and heating loads*, each having a demand factor associated with the types of loads connected. In multifamily dwellings, the total connected load should be calculated without demand factors applied, then a demand factor will be applied based on the number of units if more than two. The first step is to determine the demand from the *multifamily dwelling calculated load*. 220.84 lists all the loads that will be part of determining the calculated load.

220.84(C)(1): 3 volt-amperes per square foot.

220.84(C)(2): All dwelling units are required to have two 20-ampere small-appliance branch circuits at 1,500 volt-amperes each, and each laundry branch circuit.

220.84(C)(3): The nameplate rating of appliances that are permanently connected or fastened in place. This includes ranges, wall-mounted ovens, cooktops, clothes dryers, water heaters, etc.

220.84(C)(4): The nameplate ampere or kVA rating of all permanently connected motors not included in item (3).

220.84(C)(5): The larger of the air-conditioning load or the fixed electric heating load. Full nameplate ratings. No demand factor.

The sum of all of these loads is the calculated load.

Calculated Load

General lighting and receptacles at 3 VA/square foot = 3,600 VA
Two small appliance branch circuits at 1,500 VA each = 3,000 VA
Laundry circuit (laundry facility located elsewhere on the premises) = 0 VA
Range = 10,000 VA
Water heater = 5,000 VA
Dryer (laundry facility located elsewhere on the premises) = 0 VA
Dishwasher = 1,200 VA
Disposal = 900 VA
Largest of air-conditioning or heating load (heat) = 5,000 VA

Total Calculated Load = 25,700 VA

Total Connected Load of Service

25,700 VA × 10 Units = 257,000 VA or 257 kVA

Application of Demand Factor (three or more units) Table 220.84

$$257,000 \times 0.43 = 110,510 \text{ VA}$$

Determining Service Size

Each unit is single-phase. However, the units are fed from a three-phase service. A three-phase formula is needed to convert volt-amperes to amperes.

$$A = \frac{(Total\ connected\ demand\ load\ in\ VA)}{(Line\ to\ line\ voltage)(1.73)}$$

$$\frac{110,510 \text{ VA}}{(208 \text{ V})(1.73)} = 307 \text{ A}$$

The calculated demand load in amperes for this multifamily dwelling is 307 amperes. Electricians need to provide, at a minimum, the calculated load of 307 amperes. This does not correspond to a standard ampere rating, so they must use the next highest rating found in 240.6(A). Thus, the service size is **350 amperes**.

58. D: 430.62(A), Table 430.52, Table 430.250. When determining the motor feeder short-circuit and ground-fault protection for motor circuit conductors sized based on 430.24, first determine the rating of the largest branch-circuit short-circuit and ground-fault protection in accordance with 430.52, then add the full-load currents of the remaining motors of the group.

Largest Rating of the Branch-Circuit Short-Circuit and Ground-Fault Protection

$$20 \text{ hp}, 460 \text{ V Motor} = 27 \text{ A per Table } 430.250$$

$27 \text{ A} \times 2.5 = 67.5 \text{ A}$

Sum of Other Motors

$$10 \text{ hp}, 460 \text{ V} = 14 \text{ A per Table } 430.250$$

$14 \text{ A} \times 2 \text{ Motors} = 28 \text{ A}$

Total

$67.5 \text{ A} + 28 \text{ A} = 95.5 \text{ A or } 100 \text{ A breaker}$

59. C: 680.22(A)(4). For pools, all 150 volt or less, 60 ampere or less receptacles installed within 20 feet of the inside wall of a pool must be protected by a Class A GFCI. This includes receptacles that are not otherwise used for pool equipment. Receptacles between 151 and 250 volts are now required to be provided with SPGFCI protection not to exceed 20-mA ground-fault trip current.

60. D: 424.22(B). The NEC requires that electric space heating equipment rated more than 48 amperes have its heating elements subdivided, and that each subdivided load be rated 48 amperes or less and comply with 424.4(B). The electric space heating in question will require two branch circuit overcurrent protection devices.

14,500 kW of total resistive heating.

$$\frac{14,500 \text{ kW}}{240 \text{ V}} = 60.42 \text{ A}$$

(This is more than 48 A, so the loads must be subdivided.)

$$\frac{11,000 \text{ kW}}{240 \text{ V}} = 45.8 \text{ A}$$

$45.8 \times 1.25 = 57.29 \text{ A}$

57.29 A or 60 amperes

$$\frac{3,500 \text{ kW}}{240 \text{ V}} = 14.58 \text{ A}$$

$14.58 \times 1.25 = 18.23 \text{ A}$

18.23 A or 20 amperes

61. A: 630.11(A) and (B). When determining the conductor ampacity for individual welders, use either the I_{1eff} on the nameplate, or the primary current and duty cycle multiplied by the factor given in Table 630.11(A). When determining the conductor ampacity for a group of welders, use the individual ampacities determined in 630.11(A), using 100% of the ampacity of the two largest welders, plus 85% of the ampacity of the third largest welder, plus 70% of the fourth largest welder, plus 60% of all remaining welders.

I_{1eff} of two largest welders = 34 amperes

Third largest welder = 23 amperes × 0.78 = 17.94 amperes.

Ampacity of welder 1 = 34 A × 100% = 34 A
Ampacity of welder 2 = 34 A × 100% = 34 A
Ampacity of welder 3 = 17.94 A × 85 % = 15.25 A
Total ampacity for the group = 83.25 A

62. C: 230.90(A) Exception No. 3 states that combined rating of the two to six circuit breakers or sets of fuses shall be permitted to exceed the ampacity of the service conductors, provided the calculated load does not exceed the ampacity of the service conductors. Therefore, service conductors must be sized to meet or exceed the 185-ampere calculated load but are not required to size the conductors according to the combined rating of the five overcurrent devices making up the service disconnecting means. As is usual, Table 310.16 is used for determining the allowable ampacity of not more than three current-carrying conductors in a raceway. Using the 75 °C column, as not to exceed the temperature limitation of most terminations, the minimum size conductor is found to be 3/0 AWG.

63. A: 320.30(B). Type AC cable (often called by its trade name, BX cable) shall be secured within 12 inches or every outlet box, junction box, cabinet, or fitting and at intervals not exceeding 4½ feet. Type AC cables are sometimes mistaken for Type MC cables. One of the main distinguishing characteristics of Type AC cable is the presence of a small internal bonding strip of copper or aluminum in contact with the armor for its entire length. The combination of the bonding strip and the armor is recognized as an equipment grounding conductor by 250.118.

64. A: 230.56 requires that the phase conductor of a four-wire delta connected service having a higher voltage to ground be marked with an outer finish that is orange in color. This type of service is often referred to as a *high-leg delta*. One phase to ground will be between 210 volts and 220 volts referenced to ground. Unlike a four-wire wye connected service, the phase-to-phase voltages are 240 volts as opposed to 208 volts. Without the appropriate markings, the potential exists for personnel and equipment to be damaged by the unknowing connection to this elevated voltage.

65. B: 314.23(E). Boxes that are supported by raceways must have integral threaded hubs and must be supported by more than one conduit, except in the case of certain wiring methods covered by the exception in 314.23(E). For all other cases, any enclosures that are supported by conduits must not exceed 100 cubic inches in size.

66. B: 440.14. Except for the special condition explained in Exception No. 1 of 440.14, air conditioning and refrigeration equipment must be provided with a disconnecting means that is both within sight and readily accessible. This is sometimes referred to as a maintenance disconnect, as it is often used by maintenance workers to ensure that energy can be isolated at the equipment and within sight of those doing the maintenance work.

67. A: 410.122. Luminaires that accept incandescent lamps are required to be permanently marked with the maximum allowable wattage of the lamps. This marking must be in lettering at least ¼-inch high. The true intent of highlighting this code section, however, is to stress the importance of leaving these markings in place. Often seen as an eyesore, these markings are frequently removed by installers or future users. Installing lamps of higher wattages can overload the luminaire's conductors, particularly in the case of a luminaire with many lampholders, such as chandeliers.

68. B: 250.30(A)(2) states that supply-side bonding jumpers of the wire type shall comply with 250.102(C) when a separately derived system and the first means of disconnect are located in separate enclosures. Using table 250.102(C)(1) reveals that a 6 AWG copper supply-side bonding jumper is required when the ungrounded conductors are 1 AWG or 1/0 AWG.

69. D: 422.16(B)(1). It is permissible to install a flexible cord on an in-sink waste disposer, provided the manufacturer's instructions allow it. Additionally, the cord must be between 18 and 36 inches long and have an attachment plug of the grounding type. The receptacle supplying the disposer must be accessible and located to protect against physical damage to the cord.

70. C: 334.10 and 334.12(A). There are many installations where the use of Type NM and NMC cable is prohibited. Where Type NM/NMC cable is used in Type I and Type II construction, it is required to be installed in a raceway approved for installation in Type I or II construction, and must not violate any other section of the NEC. Type I and II buildings differ from other types of construction due to the requirement that all structural elements be non-combustible. If any part of a structure is made of wood, it is most likely a Type III, IV, or V, and therefore would permit the use of NM/NMC cable.

71. D: 230.50(B)(1), 352.10(F). Where subject to physical damage, underground service entrance conductors must be protected by an appropriate raceway. While Schedule 40 PVC is acceptable for exposed use, schedule 80 PVC is required when subject to physical damage.

72. D: 314.28(A)(1). Pull boxes containing conductors larger than 4 AWG and required to be insulated must be sized according to 314.28(A). For straight pulls, the length of the box must be at least eight times the diameter of the largest raceway. In this question, the largest raceway is the 4-inch RMC.

4 inches $\times$ 8 $=$ 32 inches

73. B: 210.70(B). A lighting outlet is required in every habitable room and bathroom of hotels and other similar occupancies. The control for these lighting outlets is permitted to be a listed wall-mounted control device. In rooms other than bathrooms and kitchens, one or more receptacles controlled by a wall-mounted control device are allowed in lieu of lighting outlets. Occupancy sensors are allowed when they supplement a listed wall-mounted control device or if they are of the wall-mounted type and are equipped with a manual override. Lighting outlets whose sole control is a pull chain are not permitted.

74. A: 430.6(A)(1), 430.110. A device serving solely as a disconnecting means for a motor must have an ampere rating not less than 115% of the full-load current of the motor. 430.6(A)(1) dictates that, when calculating the ratings of disconnecting switches, the values calculated in Table 430.247, Table 430.248, Table 430.249, or Table 430.250 should be used.

75. A: Table 450.3(B). When sizing overcurrent protection for transformers 1,000 volts and less, Table 450.3(B) is used. The first step is to determine the transformers rated current for the primary and, if necessary, the secondary of the transformer. Next, apply the percentages based on the protection method and the appropriate column for the rated current.

Rated Current of Primary

$$\frac{1,000 \text{ VA}}{480} = 2.08 \text{ A}$$

For primary protection only with currents of more than 2 amperes but less than 9 amperes, the percentage multiplier is 167%.

2.08 A × 1.67 = 3.47 A, or the next highest standard rating of 6 amperes.

Note 1 of Table 450.3(B) states that when this current does not correspond to a standard rating of a fuse or nonadjustable circuit breaker, a higher rating that does not exceed the next higher standard rating shall be permitted. Standard overcurrent ratings can be found in 240.6(A).

76. B: Power factor is the ratio of "true" (or "real") power to the "apparent" power flowing in a circuit. The discrepancy between the true power in a circuit and its apparent power is due to the voltage and current being out-of-phase. This happens when reactive loads, such as those having large inductive or capacitive components, are added to a system. The most common source of poor power factor is inductive motor loads. Inductive loads, such as motors, generate a counter-electromotive force that opposes a change in current in the electromotive force that induced it. This causes the peak current to lag the peak voltage. The amount of lag between the current and the voltage is directly proportional to the power factor. On the other hand, capacitive reactance will tend to oppose a change in voltage, thus bringing peak current and voltage closer together. The most common solution to the problem of poor power factor due to a large number of inductive loads is to add power factor correction capacitors. Motors that are lightly loaded tend to be the largest contributors to poor power factor. VFDs are a good solution to power factor correction for these types of loads, however they are less effective if the motor is already operating near its full load amperage.

77. D: 215.10 Exception No. 1. While it is almost always safer to install ground-fault protection of equipment on feeders and services rated 1,000 amperes or more than 150 volts phase-to-ground (480/277V wye), the installation of such protection on continuous industrial processes where a nonorderly shutdown would introduce additional hazards is not required. Additionally, this also applies if the supply side of the feeders is protected by a ground-fault protection.

78. C: The turns ratio of a transformer determines the difference between the voltage supplied to the primary of a transformer and the resulting voltage that is induced on the secondary. The transformer in this question has a turns ratio of 4-to-1.

$$\frac{Primary\ turns}{Secondary\ turns} = \frac{1,200}{300} = \frac{4}{1}$$

The voltage induced on the secondary is equal to the voltage on the primary and divided by the turns ratio.

$$\frac{480\ V}{\frac{4}{1}} = \frac{480\ V}{4} = 120\ volts$$

79. B: 230.80 requires that the combined ratings of all disconnects meet the requirements of 230.79. Though 230.79 stipulates that the disconnecting means shall have a rating no less than the calculated load, it also requires that the minimum rating for a one-family dwelling not be less than 100 amperes, meaning the minimum combined rating of the service disconnecting means.

80. A: 215.12(C)(2). Ungrounded conductors sized number 6 AWG or smaller are generally not allowed to be identified simply with marking tape. Instead, they must be of the appropriate continuous outer finish or be permanently marked in the other appropriate manners.

81. A: 314.24(A). Outlet boxes that do not enclose devices or utilization equipment shall have a minimum internal depth of ½ inch. For certain boxes, such as pancake boxes designed to be installed flush with a ceiling support, a ½-inch depth is required so as to not extend past the finish, i.e., the drywall. Where these boxes support a device or fixture, commonly a ceiling fan, the device must have a raised cover to allow room for wiring.

82. A: Table 314.16(A), Table 314.16(B), 314.16(B)(1) through (5). To determine the number of additional conductors that can be pulled through a box, electricians must first determine the volume of the conductors and devices already installed in the box.

Volume of Existing Components

Two 12 AWG Conductors	= 4.5 cubic inches [314.16(B)(1)] [Table 314.16(B)]
One 12 AWG equipment grounding conductor	= 2.25 cubic inches [314.16(B)(5)] [Table 314.16(B)] [2]
One receptacle	= 4.5 × cubic inches [314.16(B)(4)] [Table 314.16(B)] [1]
Total volume in box	= 11.25 cubic inches

1: 314.16(B)(4) – For each yoke or strap containing a double volume, allowance in accordance with Table 314.16(B) shall be made for the largest conductor connected.

2: 314.16(B)(5) – For up to four equipment grounding/bonding conductors, a single-volume allowance in accordance with Table 314.16(B) with ¼-volume allowance for every EGC thereafter.

The total volume occupied by the existing components is 11¼ cubic inches. To find the remaining volume, simply subtract the existing volume from the volume listed in Table 314.16(A) for a 4-inch by a 1½-inch box.

21 cubic inches − 11.25 cubic inches = 9.75 cubic inches

The remaining volume of 9.75 cubic inches allows for the installation of four additional 12 AWG conductors.

$$\frac{9.75 \text{ cubic inches}}{2.25 \text{ cubic inches}} = 4.333 \text{ or 4 conductors}$$

83. C: 455.8(C)(1). Phase converters are devices that convert single-phase electrical power to three-phase. Currents on the single-phase portion of the converter will be influenced by the three-phase load that is connected to the converter. 455.8(C)(1) requires that the disconnecting means on the single-phase portion be 250% of the full-load amperage of the three-phase motor(s) and the sum of any other loads served.

60 A × 2.5 = 150 A

84. B: 520.41(A). Footlights, border lights, and proscenium sidelights shall be arranged so that no branch circuit supplying such equipment carries a load exceeding 20 amperes.

85. C: 314.16(A), 314.16(B)(1) through (5), Table 314.16(A), Table 314.16(B). This question requires the box to be sized according to a given box fill calculated by adding up the conductor fill, clamp fill, device fill, and equipment ground conductor fill.

Volume of Components

Two 12 AWG conductors	= 4.5 cubic inches [Table 314.16(B)]
Two 10 AWG conductors	= 5 cubic inches [Table 314.16(B)]
One receptacle for 12 AWG conductors	= 4.5 cubic inches [Table 314.16(B)] [1]
One receptacle for 10 AWG conductors	= 5 cubic inches [Table 314.16(B)] [1]
One 10 AWG equipment grounding conductor	= 2.5 cubic inches [Table 314.16(B)] [2]
One internal cable clamp	= 2.5 cubic inches [Table 314.16(B)] [3]
Total volume allowance	= 24 cubic inches

1: 314.16(B)(4) – For each yoke or strap containing a double volume, allowance in accordance with Table 314.16(B) shall be made for the largest conductor connected.

2: 314.16(B)(5) – For up to four equipment grounding/bonding conductors, a single-volume allowance in accordance with Table 314.16(B) and ¼-volume allowance for every EGC thereafter.

3: 314.16(B)(3) – Where one or more internal cable clamps are present

86. D: When installing receptacles for the installation of specific equipment, 210.50(C) requires that the receptacle be installed within 6 feet of the intended location of the appliance.

87. C: 430.6(A)(1), Table 430.250, Table 430.52. When sizing motor branch-circuit short-circuit and ground-fault protection for three-phase AC induction motors, Table 230.250 is used to determine the current that will be applied to Table 430.52 based on the horsepower and rated voltage of the motor. The current listed for a 20 horsepower, 230-volt, three-phase motor is 54 amperes.

Table 430.52 gives the maximum rating or setting for the motor branch-circuit short-circuit and ground-fault protection as a percentage of the current from Table 430.248.

$$54\,A \times 2.5 = 135\,A$$

Exception No. 1 Table 430.52 allows the next highest standard ampere rating for a protection device to be used when the calculated rating does not correspond to a standard ampere rating. In this case, 135 amperes do not correspond to a standard ampere rating, and the next highest standard rating is 150 amperes.

Note: *Motor branch-circuit short-circuit and ground-fault protection* must be sized large enough to prevent nuisance tripping during startup due to inrush current, but still protect against line-to-line and line-to-ground faults within the motor and motor circuit. *Motor overload protection* protects against overloads above the designed limitations of the motor and motor circuit conductors but below the ratings of the *branch-circuit short-circuit and ground-fault protection.*

88. A: Table 310.16 is used for determining the ampacity of insulated conductors with no more than three current-carrying conductors in a raceway, cable, or directly buried. Table 310.16 is also based on an ambient temperature of 30 °C (86 °F). When these conditions change, the ampacity of the conductors must be adjusted in accordance with 310.15(B) and (C). Temperature corrections and adjustment factors shall be permitted to be applied to the ampacity for the temperature rating of the conductor. Though electricians often base conductor sizing off of the temperature limitations of terminations and equipment, when adjusting for temperature or conductor bundling, they are permitted to do the adjustment based on the conductor temperature rating, i.e., the 90 °C column for THHN.

The conditions of use for the conductors in this question require that the ampacity be adjusted to compensate for both the increased ambient temperature beyond the design of Table 310.16 and as the adjustment required for more than three current-carrying conductors. Table 310.15(B)(1) is used to determine the ambient temperature adjustment factor when the ambient temperature is not 30 °C. Additionally, 310.15(B)(2) requires that, for raceways and cables installed within ⅞ inches of a rooftop, measured from the bottom of the raceway to the roof surface, it is required to add 33 °C to the ambient temperature to determine the adjustment factor. With the 33 °C added to the 38 °C ambient temperature for a combined temperature of 71 °C, it reveals an ambient temperature correction factor of 0.5 °C. Next, Table 310.15(C)(1) is used to determine the adjustment factor when more than three current-carrying conductors are installed together. Using Table 310.15(C)(1), an adjustment factor of 80% must be applied to the ampacity of the conductors. Based on Table 310.16, the ampacity of 12 AWG copper THHW is 30 amperes.

$$30\ A \times 0.5 = 15\ A$$

$$15\ A \times 0.8 = 12\ A$$

89. D: 314.28(E). In general, power distribution blocks shall not be used in pull or junction boxes smaller than 100 cubic inches. When the distribution block is being used for the connection of equipment grounding conductors, it is permitted to be installed in smaller enclosures.

90. D: 445.19(A), 445.19(B). Generators larger than 15 kW must be provided with a remote stop switch located outside the equipment room or generator enclosure. This must also be equipped with provisions to shut down all prime mover start control circuits and initiate a shutdown mechanism that requires a mechanical reset.

91. B: 240.40. A disconnecting means must be provided on the supply side of all cartridge fuses and all fuses over 150 volts to ground. Replacing energized fuses poses an obvious risk of electrical shock, but can also result in maintenance personnel being exposed to arcing fault current when the initial cause of the fuse blowing was due to a fault.

92. B: 334.30. Nonmetallic-sheathed cable should be secured by staples, listed cable ties, straps, hangers, or similar fittings at intervals not exceeding 4½ feet and within 12 inches of every cable entry in outlet boxes. Sometimes installers form cables into boxes or extend cables laterally within a framing bay before the cables enter a box. Even when a cable is secured within 12 inches of a box, the length of the actual cable must not be longer than 18 inches between the cable entry and the cable's closest support.

93. D: 691.4. Out of all the requirements listed in 691.4 for large-scale PV electric supply stations, a requirement to limit voltages to 50 kV or less is not one of them. Interconnections to the utility must be done through medium- or high-voltage switch gear, etc. Many high-voltage transmission lines are well above 50 kV.

94. D: 210.12. When using an outlet branch-circuit-type arc-fault circuit interrupter, any of the answers listed are sufficient to provide the necessary protection along the entire length of the circuit.

95. A: 210.52(C)(3). The 2020 NEC allowed receptacle outlets to be installed below countertops as long as they were within 12 inches of the surface and not installed where the countertop extended more than 6 inches from its support base. The 2023 NEC has removed this option, leaving only above or within the countertop as possible installation locations.

96. C: 314.27(A)(2). Any luminaire weighing more than 50 pounds must have a supporting means independent of the outlet box. Boxes designed for the support of luminaires weighing more than 50 pounds must be listed and marked with the maximum weight.

97. D: 230.95. For services of more than 150 volts to ground and rated more than 1,000 amperes, such as large capacity 480Y/277-volt solidly grounded services, ground-fault protection of equipment is required to protect against catastrophic failure due to line-to-ground faults and large available fault currents.

98. A: 314.28(A)(2). For angle, U pull, or splice boxes containing 4 AWG insulated conductors, the minimum distance between raceways enclosing the same conductors must be six times the diameter of the larger raceway. If two 2-inch conduits are entering the same box on adjacent walls and contain the same conductors, the minimum straight-line distance between them must be at least 12 inches.

99. B: 210.62 covers the requirements for receptacles installed in show windows. At least one receptacle must be installed within 18 inches of the top of each show window. No point along the top of the window may be further than 6 feet from a receptacle outlet. This particular window would therefore need two receptacles.

100. D: 330.30(D). Where fished in concealed spaces and securing is impractical, when the cable is not more than 6 feet in length from its last point of support to point of connection to lights or equipment in an accessible ceiling, or when flexibility is required and not more than 3 feet from the last point where it is securely fastened, Type MC cable is permitted to be unsupported and unsecured beyond 12 inches for securing at outlet boxes, and supporting at not more than 6-foot intervals.

How to Overcome Test Anxiety

Just the thought of taking a test is enough to make most people a little nervous. A test is an important event that can have a long-term impact on your future, so it's important to take it seriously and it's natural to feel anxious about performing well. But just because anxiety is normal, that doesn't mean that it's helpful in test taking, or that you should simply accept it as part of your life. Anxiety can have a variety of effects. These effects can be mild, like making you feel slightly nervous, or severe, like blocking your ability to focus or remember even a simple detail.

If you experience test anxiety—whether severe or mild—it's important to know how to beat it. To discover this, first you need to understand what causes test anxiety.

Causes of Test Anxiety

While we often think of anxiety as an uncontrollable emotional state, it can actually be caused by simple, practical things. One of the most common causes of test anxiety is that a person does not feel adequately prepared for their test. This feeling can be the result of many different issues such as poor study habits or lack of organization, but the most common culprit is time management. Starting to study too late, failing to organize your study time to cover all of the material, or being distracted while you study will mean that you're not well prepared for the test. This may lead to cramming the night before, which will cause you to be physically and mentally exhausted for the test. Poor time management also contributes to feelings of stress, fear, and hopelessness as you realize you are not well prepared but don't know what to do about it.

Other times, test anxiety is not related to your preparation for the test but comes from unresolved fear. This may be a past failure on a test, or poor performance on tests in general. It may come from comparing yourself to others who seem to be performing better or from the stress of living up to expectations. Anxiety may be driven by fears of the future—how failure on this test would affect your educational and career goals. These fears are often completely irrational, but they can still negatively impact your test performance.

> **Review Video: <u>3 Reasons You Have Test Anxiety</u>**
> Visit mometrix.com/academy and enter code: 428468

Elements of Test Anxiety

As mentioned earlier, test anxiety is considered to be an emotional state, but it has physical and mental components as well. Sometimes you may not even realize that you are suffering from test anxiety until you notice the physical symptoms. These can include trembling hands, rapid heartbeat, sweating, nausea, and tense muscles. Extreme anxiety may lead to fainting or vomiting. Obviously, any of these symptoms can have a negative impact on testing. It is important to recognize them as soon as they begin to occur so that you can address the problem before it damages your performance.

> **Review Video: 3 Ways to Tell You Have Test Anxiety**
> Visit mometrix.com/academy and enter code: 927847

The mental components of test anxiety include trouble focusing and inability to remember learned information. During a test, your mind is on high alert, which can help you recall information and stay focused for an extended period of time. However, anxiety interferes with your mind's natural processes, causing you to blank out, even on the questions you know well. The strain of testing during anxiety makes it difficult to stay focused, especially on a test that may take several hours. Extreme anxiety can take a huge mental toll, making it difficult not only to recall test information but even to understand the test questions or pull your thoughts together.

> **Review Video: How Test Anxiety Affects Memory**
> Visit mometrix.com/academy and enter code: 609003

Effects of Test Anxiety

Test anxiety is like a disease—if left untreated, it will get progressively worse. Anxiety leads to poor performance, and this reinforces the feelings of fear and failure, which in turn lead to poor performances on subsequent tests. It can grow from a mild nervousness to a crippling condition. If allowed to progress, test anxiety can have a big impact on your schooling, and consequently on your future.

Test anxiety can spread to other parts of your life. Anxiety on tests can become anxiety in any stressful situation, and blanking on a test can turn into panicking in a job situation. But fortunately, you don't have to let anxiety rule your testing and determine your grades. There are a number of relatively simple steps you can take to move past anxiety and function normally on a test and in the rest of life.

> **Review Video: How Test Anxiety Impacts Your Grades**
> Visit mometrix.com/academy and enter code: 939819

Physical Steps for Beating Test Anxiety

While test anxiety is a serious problem, the good news is that it can be overcome. It doesn't have to control your ability to think and remember information. While it may take time, you can begin taking steps today to beat anxiety.

Just as your first hint that you may be struggling with anxiety comes from the physical symptoms, the first step to treating it is also physical. Rest is crucial for having a clear, strong mind. If you are tired, it is much easier to give in to anxiety. But if you establish good sleep habits, your body and mind will be ready to perform optimally, without the strain of exhaustion. Additionally, sleeping well helps you to retain information better, so you're more likely to recall the answers when you see the test questions.

Getting good sleep means more than going to bed on time. It's important to allow your brain time to relax. Take study breaks from time to time so it doesn't get overworked, and don't study right before bed. Take time to rest your mind before trying to rest your body, or you may find it difficult to fall asleep.

> **Review Video: <u>The Importance of Sleep for Your Brain</u>**
> Visit mometrix.com/academy and enter code: 319338

Along with sleep, other aspects of physical health are important in preparing for a test. Good nutrition is vital for good brain function. Sugary foods and drinks may give a burst of energy but this burst is followed by a crash, both physically and emotionally. Instead, fuel your body with protein and vitamin-rich foods.

Also, drink plenty of water. Dehydration can lead to headaches and exhaustion, especially if your brain is already under stress from the rigors of the test. Particularly if your test is a long one, drink water during the breaks. And if possible, take an energy-boosting snack to eat between sections.

> **Review Video: <u>How Diet Can Affect your Mood</u>**
> Visit mometrix.com/academy and enter code: 624317

Along with sleep and diet, a third important part of physical health is exercise. Maintaining a steady workout schedule is helpful, but even taking 5-minute study breaks to walk can help get your blood pumping faster and clear your head. Exercise also releases endorphins, which contribute to a positive feeling and can help combat test anxiety.

When you nurture your physical health, you are also contributing to your mental health. If your body is healthy, your mind is much more likely to be healthy as well. So take time to rest, nourish your body with healthy food and water, and get moving as much as possible. Taking these physical steps will make you stronger and more able to take the mental steps necessary to overcome test anxiety.

Mental Steps for Beating Test Anxiety

Working on the mental side of test anxiety can be more challenging, but as with the physical side, there are clear steps you can take to overcome it. As mentioned earlier, test anxiety often stems from lack of preparation, so the obvious solution is to prepare for the test. Effective studying may be the most important weapon you have for beating test anxiety, but you can and should employ several other mental tools to combat fear.

First, boost your confidence by reminding yourself of past success—tests or projects that you aced. If you're putting as much effort into preparing for this test as you did for those, there's no reason you should expect to fail here. Work hard to prepare; then trust your preparation.

Second, surround yourself with encouraging people. It can be helpful to find a study group, but be sure that the people you're around will encourage a positive attitude. If you spend time with others who are anxious or cynical, this will only contribute to your own anxiety. Look for others who are motivated to study hard from a desire to succeed, not from a fear of failure.

Third, reward yourself. A test is physically and mentally tiring, even without anxiety, and it can be helpful to have something to look forward to. Plan an activity following the test, regardless of the outcome, such as going to a movie or getting ice cream.

When you are taking the test, if you find yourself beginning to feel anxious, remind yourself that you know the material. Visualize successfully completing the test. Then take a few deep, relaxing breaths and return to it. Work through the questions carefully but with confidence, knowing that you are capable of succeeding.

Developing a healthy mental approach to test taking will also aid in other areas of life. Test anxiety affects more than just the actual test—it can be damaging to your mental health and even contribute to depression. It's important to beat test anxiety before it becomes a problem for more than testing.

> **Review Video: Test Anxiety and Depression**
> Visit mometrix.com/academy and enter code: 904704

Study Strategy

Being prepared for the test is necessary to combat anxiety, but what does being prepared look like? You may study for hours on end and still not feel prepared. What you need is a strategy for test prep. The next few pages outline our recommended steps to help you plan out and conquer the challenge of preparation.

STEP 1: SCOPE OUT THE TEST

Learn everything you can about the format (multiple choice, essay, etc.) and what will be on the test. Gather any study materials, course outlines, or sample exams that may be available. Not only will this help you to prepare, but knowing what to expect can help to alleviate test anxiety.

STEP 2: MAP OUT THE MATERIAL

Look through the textbook or study guide and make note of how many chapters or sections it has. Then divide these over the time you have. For example, if a book has 15 chapters and you have five days to study, you need to cover three chapters each day. Even better, if you have the time, leave an extra day at the end for overall review after you have gone through the material in depth.

If time is limited, you may need to prioritize the material. Look through it and make note of which sections you think you already have a good grasp on, and which need review. While you are studying, skim quickly through the familiar sections and take more time on the challenging parts. Write out your plan so you don't get lost as you go. Having a written plan also helps you feel more in control of the study, so anxiety is less likely to arise from feeling overwhelmed at the amount to cover.

STEP 3: GATHER YOUR TOOLS

Decide what study method works best for you. Do you prefer to highlight in the book as you study and then go back over the highlighted portions? Or do you type out notes of the important information? Or is it helpful to make flashcards that you can carry with you? Assemble the pens, index cards, highlighters, post-it notes, and any other materials you may need so you won't be distracted by getting up to find things while you study.

If you're having a hard time retaining the information or organizing your notes, experiment with different methods. For example, try color-coding by subject with colored pens, highlighters, or post-it notes. If you learn better by hearing, try recording yourself reading your notes so you can listen while in the car, working out, or simply sitting at your desk. Ask a friend to quiz you from your flashcards, or try teaching someone the material to solidify it in your mind.

STEP 4: CREATE YOUR ENVIRONMENT

It's important to avoid distractions while you study. This includes both the obvious distractions like visitors and the subtle distractions like an uncomfortable chair (or a too-comfortable couch that makes you want to fall asleep). Set up the best study environment possible: good lighting and a comfortable work area. If background music helps you focus, you may want to turn it on, but otherwise keep the room quiet. If you are using a computer to take notes, be sure you don't have any other windows open, especially applications like social media, games, or anything else that could distract you. Silence your phone and turn off notifications. Be sure to keep water close by so you stay hydrated while you study (but avoid unhealthy drinks and snacks).

Also, take into account the best time of day to study. Are you freshest first thing in the morning? Try to set aside some time then to work through the material. Is your mind clearer in the afternoon or evening? Schedule your study session then. Another method is to study at the same time of day that you will take the test, so that your brain gets used to working on the material at that time and will be ready to focus at test time.

STEP 5: STUDY!

Once you have done all the study preparation, it's time to settle into the actual studying. Sit down, take a few moments to settle your mind so you can focus, and begin to follow your study plan. Don't give in to distractions or let yourself procrastinate. This is your time to prepare so you'll be ready to fearlessly approach the test. Make the most of the time and stay focused.

Of course, you don't want to burn out. If you study too long you may find that you're not retaining the information very well. Take regular study breaks. For example, taking five minutes out of every hour to walk briskly, breathing deeply and swinging your arms, can help your mind stay fresh.

As you get to the end of each chapter or section, it's a good idea to do a quick review. Remind yourself of what you learned and work on any difficult parts. When you feel that you've mastered the material, move on to the next part. At the end of your study session, briefly skim through your notes again.

But while review is helpful, cramming last minute is NOT. If at all possible, work ahead so that you won't need to fit all your study into the last day. Cramming overloads your brain with more information than it can process and retain, and your tired mind may struggle to recall even previously learned information when it is overwhelmed with last-minute study. Also, the urgent nature of cramming and the stress placed on your brain contribute to anxiety. You'll be more likely to go to the test feeling unprepared and having trouble thinking clearly.

So don't cram, and don't stay up late before the test, even just to review your notes at a leisurely pace. Your brain needs rest more than it needs to go over the information again. In fact, plan to finish your studies by noon or early afternoon the day before the test. Give your brain the rest of the day to relax or focus on other things, and get a good night's sleep. Then you will be fresh for the test and better able to recall what you've studied.

STEP 6: TAKE A PRACTICE TEST

Many courses offer sample tests, either online or in the study materials. This is an excellent resource to check whether you have mastered the material, as well as to prepare for the test format and environment.

Check the test format ahead of time: the number of questions, the type (multiple choice, free response, etc.), and the time limit. Then create a plan for working through them. For example, if you have 30 minutes to take a 60-question test, your limit is 30 seconds per question. Spend less time on the questions you know well so that you can take more time on the difficult ones.

If you have time to take several practice tests, take the first one open book, with no time limit. Work through the questions at your own pace and make sure you fully understand them. Gradually work up to taking a test under test conditions: sit at a desk with all study materials put away and set a timer. Pace yourself to make sure you finish the test with time to spare and go back to check your answers if you have time.

After each test, check your answers. On the questions you missed, be sure you understand why you missed them. Did you misread the question (tests can use tricky wording)? Did you forget the information? Or was it something you hadn't learned? Go back and study any shaky areas that the practice tests reveal.

Taking these tests not only helps with your grade, but also aids in combating test anxiety. If you're already used to the test conditions, you're less likely to worry about it, and working through tests until you're scoring well gives you a confidence boost. Go through the practice tests until you feel comfortable, and then you can go into the test knowing that you're ready for it.

Test Tips

On test day, you should be confident, knowing that you've prepared well and are ready to answer the questions. But aside from preparation, there are several test day strategies you can employ to maximize your performance.

First, as stated before, get a good night's sleep the night before the test (and for several nights before that, if possible). Go into the test with a fresh, alert mind rather than staying up late to study.

Try not to change too much about your normal routine on the day of the test. It's important to eat a nutritious breakfast, but if you normally don't eat breakfast at all, consider eating just a protein bar. If you're a coffee drinker, go ahead and have your normal coffee. Just make sure you time it so that the caffeine doesn't wear off right in the middle of your test. Avoid sugary beverages, and drink enough water to stay hydrated but not so much that you need a restroom break 10 minutes into the test. If your test isn't first thing in the morning, consider going for a walk or doing a light workout before the test to get your blood flowing.

Allow yourself enough time to get ready, and leave for the test with plenty of time to spare so you won't have the anxiety of scrambling to arrive in time. Another reason to be early is to select a good seat. It's helpful to sit away from doors and windows, which can be distracting. Find a good seat, get out your supplies, and settle your mind before the test begins.

When the test begins, start by going over the instructions carefully, even if you already know what to expect. Make sure you avoid any careless mistakes by following the directions.

Then begin working through the questions, pacing yourself as you've practiced. If you're not sure on an answer, don't spend too much time on it, and don't let it shake your confidence. Either skip it and come back later, or eliminate as many wrong answers as possible and guess among the remaining ones. Don't dwell on these questions as you continue—put them out of your mind and focus on what lies ahead.

Be sure to read all of the answer choices, even if you're sure the first one is the right answer. Sometimes you'll find a better one if you keep reading. But don't second-guess yourself if you do immediately know the answer. Your gut instinct is usually right. Don't let test anxiety rob you of the information you know.

If you have time at the end of the test (and if the test format allows), go back and review your answers. Be cautious about changing any, since your first instinct tends to be correct, but make sure you didn't misread any of the questions or accidentally mark the wrong answer choice. Look over any you skipped and make an educated guess.

At the end, leave the test feeling confident. You've done your best, so don't waste time worrying about your performance or wishing you could change anything. Instead, celebrate the successful completion of this test. And finally, use this test to learn how to deal with anxiety even better next time.

Review Video: 5 Tips to Beat Test Anxiety
Visit mometrix.com/academy and enter code: 570656

Important Qualification

Not all anxiety is created equal. If your test anxiety is causing major issues in your life beyond the classroom or testing center, or if you are experiencing troubling physical symptoms related to your anxiety, it may be a sign of a serious physiological or psychological condition. If this sounds like your situation, we strongly encourage you to seek professional help.

Thank You

We at Mometrix would like to extend our heartfelt thanks to you, our friend and patron, for allowing us to play a part in your journey. It is a privilege to serve people from all walks of life who are unified in their commitment to building the best future they can for themselves.

The preparation you devote to these important testing milestones may be the most valuable educational opportunity you have for making a real difference in your life. We encourage you to put your heart into it—that feeling of succeeding, overcoming, and yes, conquering will be well worth the hours you've invested.

We want to hear your story, your struggles and your successes, and if you see any opportunities for us to improve our materials so we can help others even more effectively in the future, please share that with us as well. **The team at Mometrix would be absolutely thrilled to hear from you!** So please, send us an email (support@mometrix.com) and let's stay in touch.

Additional Bonus Material

Due to our efforts to try to keep this book to a manageable length, we've created a link that will give you access to all of your additional bonus material:

mometrix.com/bonus948/electricianjou

Made in the USA
Las Vegas, NV
01 March 2024

86586386R00070